Edgar Richards

Principles and Methods of Soil Analysis

Edgar Richards

Principles and Methods of Soil Analysis

ISBN/EAN: 9783743320369

Manufactured in Europe, USA, Canada, Australia, Japa

Cover: Foto ©ninafisch / pixelio.de

Manufactured and distributed by brebook publishing software
(www.brebook.com)

Edgar Richards

Principles and Methods of Soil Analysis

DEPARTMENT OF AGRICULTURE.

DIVISION OF CHEMISTRY.

BULLETIN   No. 10.

PRINCIPLES AND METHODS

OF

SOIL ANALYSIS.

EDGAR RICHARDS,

ASSISTANT CHEMIST.

WASHINGTON:

GOVERNMENT PRINTING OFFICE.

1886.

13735—No. 10

LETTERS OF TRANSMITTAL

UNITED STATES DEPARTMENT OF AGRICULTURE,
DIVISION OF CHEMISTRY,
Washington, D. C., April 2, 1886.

SIR: The numerous inquiries received at this office relating to the methods and objects of soil analysis lead me to believe that an abstract of the present knowledge possessed by scientists on this subject would prove of interest to those engaged in scientific agriculture.

Quite a number of samples of soil having accumulated in the Laboratory awaiting examination, I requested Mr. Edgar Richards to conduct the analyses and to collect the information which is hereby submitted. The following report shows how well he has performed his work. The analyst who desires information concerning the latest methods of analysis will find the best authorities mentioned with references to their works; while the reader who is not a chemist will discover a valuable fund of information about soils and their treatment.

Respectfully,

H. W. WILEY,
Chemist.

Hon. NORMAN J. COLMAN,
Commissioner of Agriculture.

UNITED STATES DEPARTMENT OF AGRICULTURE,
DIVISION OF CHEMISTRY,
Washington, D. C., March 31, 1886.

SIR: In submitting the following report an attempt has been made to collect such information on the subject of soils, now scattered through so many different books and periodicals, as will be of general interest; together with the chemical methods that I have employed in executing the analyses of some soils from New York, Louisiana, and elsewhere, during the past year.

Respectfully, yours,

EDGAR RICHARDS,
Assistant Chemist.

Dr. H. W. WILEY,
Chemist.

3

TABLE OF CONTENTS.

The following works have been consulted in the preparation of this report:

Economic Geology, by David Page. Edinburgh, 1874.
Manual of Geology, by James D. Dana. New York, 1875.
Iowa Geological Survey. Vol. I, 1858.
Kentucky Geological Survey. Vol. III, 1857.
Vegetable mold and earth worms, by Charles Darwin. New York, 1882.
Tenth Census of the United States. Vol. V. Cotton production. Washington, 1884.
The Journal of the Royal Agricultural Society.
The American Journal of Science.
The Soil of the Farm, by John Scott and J. C. Morton. London, 1882.
The Chemistry of the Farm, by R. Warington. London, 1881.
Traité de Chimie Analytique appliquée a L'Agriculture, par Eug. Peligot. Paris, 1883.
Traité D'Analyse des Matières Agricoles, par L. Grandeau. Paris, 1877.
Quantitative Chemical Analysis, by Dr. C. Remigius Fresenius. Second American edition. New York, 1883.
Quantitative Chemical Analysis, by F. A. Cairns. New York, 1881.
Volumetric Analysis, by Francis Sutton. Fifth edition. London. 1882.
Select Methods in Chemical Analysis, by Wm. Crooks, London. 1871.

PRINCIPLES AND METHODS OF SOIL ANALYSIS.

ON THE DERIVATION AND THE FORMATION OF SOIL.

All soils are the result of the natural disintegration of the rocks by atmospheric agencies, mingled with decayed vegetable and animal matter in greater or less proportion. If natural agencies, such as glaciers, rain, frost, wind, &c., did not come into play and wash and transport the materials of soil to a greater or less distance from their sources, the soil of every locality would be simply the decayed upper surface of the underlying rocks. But in proportion to the slope of the ground and the activity of the agents above mentioned, the soil is transported from higher to lower levels, and in many cases a good soil may be found covering rocks which of themselves would only yield a poor soil.

COMPOSITION OF THE SOIL.

Soil is a mixture of sand, either quartzose or feldspathic, clay, carbonate of lime, and humus or organic matter, and on the preponderance of one or more of these constituents the usual classifications of soils are based.

GENERAL CLASSIFICATION OF SOILS.

Soils are usually classified as sandy, sandy or light loams, loams, clayey loams, heavy or retentive clays, marls, calcareous loams, and peaty soils. This classification has reference chiefly to composition and texture, a special chemical composition, siliceous, calcareous, &c., being necessary for the profitable growth of particular crops, and a certain mechanical texture, friable, porous, &c., suiting best for the permeation of rain and air, and the spreading of the roots of the plant.*

Loams, which may be considered as typical soils, are a mixture of sand, clay, and humus. which are spoken of as *light* when the sand predominates and as *heavy* when the clay is in excess. These terms, *light* and *heavy*, do not refer to the actual weight of the soil, but to its tenacity and degree of resistance it offers to the implements used in cultivation. Sandy soils are, in the farmer's sense of the word, the lightest of all soils because they are the easiest to work, whilst in actual weight they are the heaviest soils known. Clay, though hard to work

---

* Page's Economic Geology.

on account of its tenacity, is comparatively a light soil in weight. Peaty soils are light in both senses of the word, being loose. or porous, and having little actual weight.* (See Table III.)

GEOLOGICAL CLASSIFICATION OF SOILS.

Whatever their composition and texture, soils are, from a geological standpoint, mainly of two sorts: *soils of disintegration* and *soils of transport.* Under the former are comprehended such as arise from the waste and decay of the immediately underlying rocks, the limestones, traps, granites, and the like, together with a certain admixture of vegetable and animal débris; and which are directly influenced in their composition, texture, and drainage by the nature of the subjacent rocks from which they are derived. Under the latter are embraced all drift and alluvial materials, such as sand, shingly débris, miscellaneous silt, and clay, which have been worn from other rocks by atmospheric agencies and transported to their existing positions by winds, waters, or ancient glacial action.†

DIFFERENCE BETWEEN THE SOIL AND THE SUBSOIL.

Besides the *soils* proper, which come immediately under cultivation, there are in most places a set of *subsoils*, differing from the true soils; and which cannot be ignored. The true soils are usually of a darker color, from the larger admixture of humus, whilst the subsoils are lighter in hue, yellow, red, or bluish, from the greater preponderance of the iron oxides. The soils are more or less friable in their texture, whilst the subsoils are tougher, more compact, and more largely commingled with rubbly and stony débris. The soils are usually a little more than mere surface covering, whilst the subsoils may be many feet in thickness.†

WEATHERING OF THE ROCKS AND FORMATION OF THE SOIL.

All exposed rocks break up in course of time under the continued action of atmospheric agencies, however hard and refractory they may be; these agencies act both chemically and mechanically. The rain, owing to the absorption of carbonic acid from the atmosphere, acts chemically on the rocks by its solvent action, and also from its oxygen combining with substances not yet fully oxidized. Its mechanical action appears in its washing away the finer portions of the disintegrated rock or soil from higher to lower ground. The changes in temperature have a loosening influence by causing alternate expansion and contraction. The atmosphere itself acts chemically upon the rocks by the slow oxidization of those minerals which can absorb more oxygen, and the production of carbonates and bicarbonates whose solubility still

* The Soil of the Farm.
† Page's Economic Geology.

further aids disintegration. These disintegrating agencies are still further aided by the root-growths of plants, by the burrowing of worms and other earth-dwelling creatures, and in no small degree by the generation of organic acids, humic, crenic, &c., by organic decay.

From the hardest granites, basalts, and lavas, to the softest limestones and marls, all are undergoing this disintegration; and the soils to which they give rise will vary in depth, composition, and texture according to the softness and mineral character of the rocks and the length of time they have been subjected to these agencies.*‡

According to Darwin† the solid rocks disintegrate even in countries where it seldom rains and where there is no frost. De Koninck, a Belgian geologist, is of opinion that such disintegration may be attributed to the carbonic and nitric acids, together with the nitrates of ammonia, which are dissolved in the dew.

The rocks which weather most easily and rapidly do not always exhibit most soil, very often the reverse. A pure limestone would exhibit hardly any weathered band or soil, because the carbonic acid of the rain would almost at once dissolve and remove the particles it acts upon. Even in the case of igneous rocks, their composition may be such that those which weather the most rapidly would, likewise, show little of a weathered band, owing to the same solvent action.‡

THE SOILS FORMED BY THE DIFFERENT GEOLOGICAL FORMATIONS.

The rocks of which feldspar is one of the constituents are the origins of the clays and potash which are met with in all arable soils; feldspar is a silicate of aluminium and potassium, which on disintegration forms clay, a silicate of aluminium, and a silicate of potassium.

The primitive and igneous rocks yield soils rich in potash, and the fossiliferous rocks those rich in phosphoric acid.

THE DENUDATION OF THE SOIL.

The same agencies which form the soils are also wasting and carrying them away. During every rain storm transportation of soil goes on, as the brooks and rivers show, after heavy long-continued rains, by the yellow muddy color of their waters that they are carrying a vast quantity of sediment towards the sea. The running streams bear along the transported matter, and gradually deposit it as the current diminishes in velocity, the very finest particles being carried as long as the stream remains in motion. When a river reaches a flat or level tract and over which its waters can flow in flood with a slow motion, the suspended matter, consisting principally of sand and mud, is deposited, and constitutes the *alluvium*, or new land, formed by such deposits at the river's mouth or along its banks.‡

* Page's Economic Geology.
† Darwin's Vegetable Mold and Earth Worms, 1882, p. 235.
‡ The Soil of the Farm.

THE QUANTITY OF SOIL SWEPT AWAY BY THE RAIN REPLACED BY THE DECOMPOSITION OF THE ROCKS.

Though the soil is thus continuously washed away, still it remains nearly constant in quantity, since what is taken away by denudation is made up from other causes, and this augmentation can proceed evidently from nothing but the slow and constant disintegration of the underlying rocks. The subsoils are likewise gradually being converted into soil, and thus keep up the supply available for the nourishment of plant life. The constant tillage and plowing of the ground subjects it more readily to the weathering action than is the case with grass or other lands protected by natural vegetation.*

THE GENERAL FERTILITY OF THE SOILS DEPENDS PRINCIPALLY ON THEIR TEXTURE.

From an agricultural standpoint, the soil, which is the natural storehouse and laboratory from whence plants derive their supply of food, should present different qualities which according as they are more or less developed, exert a considerable influence upon its fertility; it should be firm enough to afford a proper degree of support for the plants that grow on it, and yet loose enough to allow the delicate fibers of the rootlets to extend themselves in all directions in search of the food of which they are in need. It must be of such a texture as to allow the free access of air, without which plants cannot live; and it must be close enough to retain, for a considerable time, the water which falls on it, and yet porous enough to allow the excess to drain away. In this respect, the nature of the subsoil and the depth of the surface soil are both important. When a soil rests immediately upon a bed of rocks or gravel, it will naturally be drier than when it rests on clay or marl. On the other hand, a clay subsoil may be of great advantage to a sandy soil, by enabling it to retain moisture longer in dry weather. These qualities depend altogether on the state of division of the soil and of its geological origin, and it is important, consequently, to study the arable soil under the two standpoints of its *physical properties* and of its *chemical nature*. (Peligot.)

THE PHYSICAL PROPERTIES OF SOILS.

The physical properties of a soil may be considered in regard to its texture, its absorbent powers, and its temperature.

Soils differ not only in chemical composition, but also in physical characteristics, the aspect, elevation, depth, climatic conditions, drainage, &c., that enter into the problem and cause the variations in the relative productiveness of two fields.

The knowledge of the inherent agricultural capabilities of the different classes of soil is still very far from being perfect, though, by the re-

* The Soil of the Farm.

searches of chemists since 1860, many important facts have been brought to light which have led to improvements in the cultivation of the land.

IMPORTANCE OF A PROPER MECHANICAL CONDITION.

It is not very difficult to adapt the plant or crop to the nature of the soil when once we know what mineral ingredients are required by the one and furnished by the other; but it demands very close observation and study and a most diligent application of means to bring the physical or mechanical properties of the soil into the state best fitted for plant growth.

The influence of mechanical operations become obvious, as the accessibility of air, moisture, and warmth, which are essential to the development of the changes that occur in the process of germination, are but slightly influenced by the chemical composition of the soil, being all dependent on its mechanical condition. And this influence is not confined to the first stages of growth and development of vegetation, but is required all through the life of the plant, for they cannot avail themselves of their full amount of food unless the state of the soil admits of the free passage of air and moisture, and is favorable to the extension of the rootlets in all directions.*

VARIATIONS IN THE TEXTURE OF SOILS INFLUENCE THEIR FERTILITY.

Soils may vary from the coarsest pebbles and loose sands to the finest and most tenacious clays. Those soils are best adapted to agriculture which consist of a mixture of sand with a moderate quantity of clay and a little vegetable matter. When the sand or other coarse material predominates, the soil is easy to till, and will grow most of the crops which are suitable to the locality; but it is deficient in the power of retaining moisture, and the soluble and volatile parts of manure. When the clay is in excess, the soil is more difficult to till, and will probably grow fewer crops, as it retains more moisture, is not easily warmed, does not admit of free access of air, and consequently does not facilitate the chemical changes in the soil and manure placed on it, which are so important to the proper nourishment of the plants.

If soils differed in nothing else than that of texture, the one which contained the greatest amount of finely divided matter would possess an advantage over those with coarser parts. One cause of this superiority consists in the greater absorptive and retentive powers which finely divided matter possesses, due probably to the immensely greater quantity of surface exposed in a given bulk or weight of the more finely divided soil.*

THE ABSORBENT AND RETENTIVE POWERS OF SOIL.

The observations of Sir H. S. Thompson,† on the absorbent and retentive powers of soil, or the power possessed by a soil to decompose

* The Soil of the Farm.
† Journal of the Royal Agricultural Society, vol. xi, p. 68.

and retain for the subsistence of the plants the ammoniacal and other salts which form the most valuable constituents of manure, and the highly important investigations of Professors Way[*] and Voelcker[†] on this subject, have had a most important bearing on practical agriculture, especially to the rational treatment and application of farm-yard manure and the economical use of artificial manures.

The investigations of Professor Way have given a new direction to the chemical study of soils, and the subject has been taken up by Liebig, Knopp, Henneberg, Stohman, Brustlein, Peters, Voelcker, Warington, and other chemists. [In the pages of the Journal of the Royal Agricultural Society of England will be found the reports of many important investigations undertaken in England to which the reader is referred for more detailed information.]

These several investigations have shown that the property of absorbing, retaining, and modifying the composition of manures belongs to every soil in a greater or less degree.

ABSORPTION OF AMMONIACAL SALTS BY VARIOUS SOILS.

The ammonia floating in the atmosphere is continually being washed into the soil, carried into it by the rains. The clay, oxide of iron, and the organic matter contained in the soils, perform the important function of absorption. This property of clay may be one of the reasons why clay lands are more suitable to wheat than are sandy soils. Although clay has this property of retaining more of these absorbed substances than sands or loams, yet it is evident that these latter soils must receive the same amount of fertilizing matter from the rains, only they have less ability for retaining or storing it up.[‡]

In regard to the absorption of ammonia and its salts by various soils, the following summary is taken from Dr. Voelcker's paper "On the chemical properties of soils:"[§]

(1) All of the soils experimented upon had the power of absorbing ammonia from its solution in water.

(2) Ammonia is never completely removed from its solution, however, weak it may be. On passing a solution of ammonia, whether weak or strong, through any kind of soil, a certain quantity of ammonia invariably passes through. No soil has the power of fixing completely the ammonia with which it is brought in contact.

(3) The absolute quantity of ammonia which is absorbed by a soil is larger when a stronger solution of ammonia is passed through it, but, relatively weaker solutions are more thoroughly exhausted than stronger ones.

(4) A soil which has absorbed as much ammonia as it will from a weak solution, takes up a fresh quantity of ammonia when it is brought into contact with a stronger solution.

(5) In passing solutions of salts of ammonia through soils, the ammonia alone is ab-

---

[*] Journ. Royal Agric. Soc., vol. xi, p. 313.
[†] Ibid., vol. xiv, p. 808.
[‡] The Soil of the Farm.
[§] Journ. Royal Agric. Soc., vol. xxi, p. 123.

sorbed, and the acids pass through, generally in combination with lime, or when lime is deficient in the soil, in combination with magnesia or other mineral bases.

(6) Soils absorb more ammonia from stronger than from weaker solutions of sulphate of ammonia, as of other ammonia salts.

(7) In no instance is the ammonia absorbed by soils from solutions of free ammonia, or from salts of ammonia, so completely or permanently fixed as to prevent water from washing out appreciable quantities of ammonia.

(8) The proportion of ammonia which is removed in the several washings is small in proportion to that retained by the soil.

(9) The power of soil to absorb ammonia from solutions of free or combined ammonia is thus greater than the power of water to redissolve it.

It may be concluded from the above that in ordinary seasons no fears need be entertained that occasional heavy rain storms will remove much ammonia from ammoniacal top-dressings, such as sulphate of ammonia, soot, guano, and similar manures, but in excessively rainy seasons, or in districts that have a large rainfall considerable quantities may be removed from land top-dressed with ammoniacal manure, even in the case of stiff clay lands.

GENERAL CONCLUSIONS IN REGARD TO THESE POWERS.

The general conclusions that may be drawn from the different investigations show that when surface waters charged with the products of vegetable decay are brought into contact with argillaceous sediment, they part to some extent with their potash, ammonia, silica, phosphoric acid, and organic matter, which remains in combination with the soil; whilst, under ordinary conditions at least, neither nitrates, soda, lime, magnesia, sulphuric acid, nor chlorine are retained. The phosphates are probably retained in combination with alumina or peroxide of iron, and the silica and organic matters enter into more or less insoluble combinations.*

The drainage waters from clay soils, especially if the soil is in a fine state of division, are found to carry off the nitrates, sulphates, chlorides, and carbonates of soda, lime, and magnesia.

THE POWER OF RETAINING MOISTURE IN THE SOIL.

The amount of moisture retained by a soil is generally in direct ratio to its contents of organic matter and its state of division. A proper degree of fineness in the particles of the soil is very important to obtain, especially if it is subjected to drought. During dry weather plants require a soil that is both retentive and absorptive of atmospheric moisture, and that soil which has this faculty will evidently raise a more vigorous crop than one without it. The materials which are most influential in soils may be arranged in the following order, when this condition of retaining moisture is considered: Organic matter, marls, clays, loams, and sands.

* The Soil of the Farm.

THE TEMPERATURE OF THE SOIL.

The temperature of a soil depends very much upon its humidity, dry lands absorbing more quickly and losing more slowly the heat than wet lands. The temperature of drained lands is consequently higher in summer than those which are undrained. The greatest difference occurs in the spring between the temperature of the atmosphere and that of the soil, as, owing to the moisture from the winter and spring storms, the soil, in consequence of the evaporation required to dry it sufficiently, but gradually acquires the proper temperature demanded by the coming vegetation. After it is once thoroughly warmed it retains a certain amount in reserve which is of benefit to the late ripening and gathering of certain crops. Dark colored soils absorb heat more rapidly than those of lighter color.*

FERTILITY OF THE SOIL DEPENDS ON CLIMATIC CONDITIONS.

In this country the soils are fertile enough, for the most part, to raise any crop desired, if the climatic conditions are favorable, and this is a point that must not be lost sight of. As it is certain that the range of the thermometer during the growing season of the year will affect the productiveness of the ground, notwithstanding a favorable composition and texture of the soil and an adequate rainfall, and disregard of such local conditions as temperature, rainfall, elevation above sea level, aspect, nearness to water, &c., will lead to very erroneous opinions of the soil. Thus, in the case of the Northwest, for example, with the severe winters and late springs and early falls, only such crops as will mature early can be raised, notwithstanding the noted fertility of its soil.

The amount of rainfall and the season of its descent determine the nature of the crops raised, and exercise a considerable influence on the fertility of the soil. The action of the rain carries the soluble ingredients which the plants require to their roots and supplies them with the necessary moisture. The soil, however, must be permeable enough to let the excess of water drain away; water-logged soils show immediate improvement when properly drained.

THE BARENNESS OF SOIL.

No soil is absolutely barren unless it contains substances poisonous to plants, such as an excess of organic acids, alkaline salts, the sulphate of iron, the sulphide of iron, or other injurious ingredients; but it may be so considered when it will not produce such crops as the farmer may wish to raise. Such a soil may, in many cases, be made productive by adding to it the constituent of which it is in need; but, if this cannot be done except at a prohibitory cost, or one at which more fertile ground can be procured, the soil may be regarded as practically worthless.

* The Soil of the Farm.

THE AVERAGE COMPOSITION OF ORDINARY FARM CROPS.

The amount of food taken from the soil by different crops is given in the following table, taken from "The Chemistry of the Farm," pp. 38, 39. This table gives the average composition of ordinary farm crops as grown in England; and the annual produce of beech, spruce, fir, and Scotch pine forests, felled for timber, the results of extensive investigations made in Bavaria.

The quantities of carbon, hydrogen, and oxygen present are omitted; also some of the smaller ash constituents. By "pure ash" is meant the ash minus sand, charcoal, and carbonic acid.

TABLE I.—*The weight and average composition of ordinary crops in pounds, per acre.*
(R. Warington.)

	Weight of crop at harvest.	Weight of crop, dry.	Total pure ash.	Nitrogen.	Sulphur.	Potash.	Soda.	Lime.	Magnesia.	Phosphoric acid.	Chlorine.	Silica.
	Lbs.	Lbs.	Lbs.	Lbs.	Lbs.	Lbs.	Lbs.	Lbs.	Lbs.	Lbs.	Lbs.	Lbs.
Wheat:												
Grain, 30 bushels...	1,800	1,530	31	33	2.7	9.7	0.9	1.0	3.7	14.3	0.2	0.5
Straw	3,158	2,653	158	12	5.1	18.2	2.5	9.2	4.0	8.4	1.7	110.6
Total crop	4,958	4,183	189	45	7.8	27.9	3.4	10.2	7.7	22.7	1.9	111.1
Barley:												
Grain, 40 bushels...	2,080	1,747	46	35	2.9	9.8	1.0	1.3	4.0	16.2	0.4	12.0
Straw	2,447	2,080	100	12	3.2	21.6	4.2	8.5	2.5	4.4	3.2	51.5
Total crop	4,527	3,827	146	47	6.1	31.4	5.2	9.8	6.5	20.6	3.6	63.5
Oats:												
Grain, 45 bushels...	1,890	1,625	54	38	3.2	8.5	1.4	2.0	3.9	11.8	......	24.8
Straw	2,835	2,353	140	14	4.8	29.6	5.9	9.8	5.3	7.1	5.5	69.3
Total crop	4,725	3,978	194	52	8.0	38.1	7.3	11.8	9.2	18.9	5.5	94.1
Meadow hay, 1½ tons*..	3,360	2,822	208	49	5.7	56.3	11.9	28.1	10.1	12.7	16.2	57.5
Red clover hay, 2 tons*.	4,480	3,763	255	102	9.4	87.4	4.1	86.1	30.9	25.1	9.4	6.8
Beans:												
Grain, 30 bushels ..	1,920	1,613	57	77	4.4	23.0	0.8	2.9	3.8	22.3	1.5	0.8
Straw, 30 bushels..	2,240	1,848	130	22	4.9	58.1	4.9	30.2	10.3	9.2	18.1	6.9
Total crop	4,160	3,461	187	99	9.3	81.1	5.7	33.1	14.1	31.5	19.6	7.7
Turnips:												
Roots, 17 tons*	38,080	3,126	218	71	15.2	108.6	17.0	25.5	5.7	22.4	10.9	2.6
Leaf, 17 tons*......	11,424	1,531	146	49	5.7	40.2	7.5	48.5	3.8	10.7	11.2	5.1
Total crop	49,504	4,657	364	120	20.9	148.8	24.5	74.0	9.5	33.1	22.1	7.7
Swedes:												
Roots, 14 tons*	31,360	3,349	163	74	14.6	63.3	22.8	19.7	6.8	16.9	6.8	3.1
Leaf, 14 tons* .. .	4,704	706	75	28	3.2	16.4	9.2	22.7	2.4	4.8	8.3	3.6
Total crop	36,064	4,055	238	102	17.8	79.7	32.0	42.4	9.2	21.7	15.1	6.7
Mangels:												
Roots, 22 tons*	49,280	5,628	410	96	4.9	191.1	75.4	24.2	19.7	34.0	40.6	16.4
Leaf	18,233	1,654	280	51	9.1	71.4	65.2	29.1	27.2	15.1	49.8	9.2
Total crop	67,513	7,282	690	147	14.0	262.5	140.6	53.3	46.9	49.1	90.4	25.6
Potatoes:												
Tubers, 6 tons*....	13,440	3,360	126	47	2.7	75.4	2.0	2.9	5.7	24.1	3.5	2.9
Haulm†	4,274	954	50	20	2.1	1.1	2.0	22.7	12.4	2.7	1.9	2.1
Total crop	17,714	4,314	176	67	4.8	76.5	4.0	25.6	18.1	26.8	5.4	5.0

*A ton of 2,240 pounds.   †Calculated from a single analysis only.

TABLE I.—*The weight and average composition of ordinary crops, &c.*—Continued.

	Weight of crop at harvest.	Weight of crop, dry.	Total pure ash.	Nitrogen.	Sulphur.	Potash.	Soda.	Lime.	Magnesia.	Phosphoric acid.	Chlorine.	Silica.
	Lbs.	Lbs.	Lbs.	Lbs.	Lbs.	Lbs.	Lbs.	Lbs.	Lbs.	Lbs.	Lbs.	Lbs.
Beech:												
Wood	....	2,822	26	....	....	4.2	0.8	12.9	3.4	2.6	......	2.2
Leaf litter	....	2,975	166	.....,.	......	8.8	1.6	73.1	10.9	9.3	......	53.9
Total produce ...	....	5,797	192	....	.....	13.0	2.4	86.0	14.3	11.9	......	56.1
Spruce fir:												
Wood	....	3,064	20	....	....	3.6	0.4	8.2	1.8	1.3	......	2.9
Leaf litter	....	2,683	121	....	....	4.3	1.5	54.4	6.2	5.7	......	44.3
Total produce ...	....	5,747	141		......	7.9	1.9	62.6	8.0	7.0	......	47.2
Scotch pine:												
Wood	....	2,884	15	....	....	2.3	0.2	9.0	1.5	1.0	......	0.5
Leaf litter	....	2,845	42	....	....	4.3	1.7	16.8	4.3	3.3	......	5.8
Total produce ...	....	5,729	57	....	......	6.6	1.9	25.8	5.8	4.3	......	6.3

From the above table we can judge of the quantity of the different soil constituents which various crops absorb from an acre of ground, and how certain plants demand some one particular ingredient more than others. In general, we may say that the cereal crops apparently possess a capacity for feeding on silicates not enjoyed by other crops, and contain a less amount of nitrogen than either the root or leguminous crops; nevertheless, they respond the most readily to nitrogenous manures. The amount of phosphoric acid is the most constant of all the constituents of crops, being concentrated in the grain. The root crops contain a large amount of potash, and are the most exhausting to the soil in consequence; they take up more nitrogen than do the cereals, besides other ash constituents, as phosphoric acid. The leguminous crops contain about twice as much nitrogen as do the cereals, and the potash and lime occur in large proportions. Silica is nearly absent. They respond most readily to potash manures.

The growth of forests is far less exhausting to a soil than are most ordinary farm crops, especially where the leaves from the trees are left to manure the ground by their decay.

PERMANENT FERTILITY.

The investigations of Messrs. Lawes and Gilbert* in regard to the exhaustion of land by the same crops grown year after year on the same field, left unmanured, which they have been carrying on at Rothamsted, England, for the past forty years, lead them to conclude that all lands left unmanured for a longer or shorter number of years have a certain

* Journ. Agric. Soc.

standard of natural fertility, varying within certain limits, according to the character of the season and the management; which standard, on a large scale, could practically neither be permanently reduced nor increased by cultivation. Such lands are said to be "out of condition."

Of course, it must be borne in mind that these observations apply to actual English farm practice and the term must not be pushed to any great extreme.

ACQUIRED OR TEMPORARY FERTILITY.

A land is said to be "in good condition" when by the application of manure its permanent fertility is raised so as to produce larger crops, due to the accumulation within the soil of suitable plant food derived from the manure, which may be reduced or entirely withdrawn by the crops. But since it is the minimum of any one essential ingredient and not the maximum of the others which is the measure of fertility, a soil may become exhausted for one plant yet still contain an abundant food supply for another plant whose food requirements are different. Thus a rotation of crops will defer the period of exhaustion. A poor soil is sooner reduced to sterility than a rich one, a shallow soil would fail sooner than a deep one, and a light soil sooner than a stiff one. As only about 1 per cent. of a soil is in a fit condition at any moment for plant food, an immense store of nourishment is contained in most soils in a passive condition, which gradually becomes available.*

IMPROVEMENT OF THE SOIL.

The improvement of the soil by tillage, drainage, irrigation, liming, and the application of manures, does not enter into the subject of this report, and the reader in quest of such information is referred to any of the standard works on agriculture, where these subjects are treated in full detail.

THE MECHANICAL ANALYSIS OF SOIL.

At one time great stress was laid upon the mechanical analysis of a soil, and chemists were told that more depended on it than on the chemical composition, but nowadays, whilst a knowledge of its physical condition is a great help in studying the nature of a soil, still its chemical analysis is of more importance.

Of the great number of apparatus proposed to effect the mechanical analysis of soils, all labor under more or less objections, and the same soil submitted to different processes yields most diverse results.

---

* The Soil of the Farm.

An Italian chemist, M. Pellegrini, obtained the following results with a clay soil of Orciano, near Pisa, on using the apparatus named: *

	Noeble's.	Schloesing's.	Masure's.
	Per cent.	*Per cent.*	*Per cent.*
Sand	1.47	32.07	13.35
Clay	87.31	37.67	71.90
Earthy carbonates	....	20.20	....
Organic and volatile matter....	9.66	10.25	....
Undetermined	....	....	14.75
Soluble and loss	1.56	......	......
	100.00	100.19	100.00

Whilst these differences are enormous, still the methods are hardly comparable. That of Schloesing's has for its object the separation of the clay, in almost a pure state, from the sand, lime, and other mate. rials which accompany it. Masure's, and Noeble's apparatus make use of the mechanical action of a stream of water to separate the soil into more or less fine particles.

OBJECTION TO THE MECHANICAL ANALYSIS OF A SOIL.

The objection most frequently urged against such mechanical analysis is, that the lightest portion, most commonly called clay, contains, in addition to that body, some very fine sand, some calcareous or feldsspathic products, in addition to organic matter in a fine state of division. This cause of error has long been pointed out by Boussingault, Gasparin, and other authors.

PRINCIPLE APPLIED TO MOST OF THE APPARATUS USED FOR THIS PURPOSE.

The principle adopted in most apparatus used for this purpose is the mechanical action of a stream of water flowing through the soil into a succession of vessels, each somewhat larger than the one preceding, and in which a certain amount of sediment is gradually deposited, beginning with the coarsest and heaviest particles and ending with the very finest. A weighed quantity of the air-dried soil is taken, and the action of the water continued till it runs through the last vessel used perfectly clear; the different deposits are collected, dried, ignited, and weighed separately. The results obtained are only approximate, and differ in the same soil using the same apparatus.

A succession of metal sieves, ranging from 10 to 100 meshes to the square inch, are sometimes used for this purpose, a weighed quantity of soil being taken and the portion remaining on each sieve being collected and weighed.

* Peligot, p. 154.

SCHLOESING'S METHOD FOR THE MECHANICAL DETERMINATION OF CLAY.

The method adopted by Schloesing for the mechanical determination of clay is as follows: Knead in a porcelain dish 5 to 10 grams of the soil, previously separated from the gravel and organic matter, with a little water into a firm paste; fill the dish half full of water, and rub the mass with the forefinger; decant without carrying over the sand; repeat the washing by decantation until the sand yields nothing further to the water. All the waters of decantation are collected together, making a volume of 300 to 400 cubic centimeters of liquid; this is treated with nitric acid in small quantities at a time until the solution is acid to test paper. This treatment has the effect of coagulating the clay and thus clearing up the muddy liquid. The solution is transferred to a filter and the mixture of fine sand and clay washed thoroughly with water until the liquid goes through cloudy; when the contents of the filter are washed with dilute ammonia into a jar of 2 liters' capacity, using about 150 c. c. of the dilute ammonia for this purpose; then fill up the jar with pure water, and let it digest for an hour, agitating it frequently. Set it aside for twenty-four hours, and then siphon off the clayey liquid; the sand resting at the bottom of the vessel is dried and weighed in a dish. This is the fine sand that is ordinarily mixed up in the clay. Its weight deducted from that of the quantity of soil used will give the clay. (Peligot, p. 153.)

THE EFFECT OF VARIOUS PROPORTIONS OF SAND IN THE SOIL.

According to Thaer (Peligot, p. 158), when the sand and clay are of equal parts, or in the proportion of 40 of sand to 60 of clay, comprising under this name the finest sand, &c., as found in mechanical analysis, the soil is fitted for all kinds of crops; with more than 60 per cent. of sand they are suitable to rye and barley, rarely for wheat; with 70 per cent. of sand, the soil is suitable still for the cultivation of barley, and especially for the cultivation of rye; it is easily worked, but the manures are rapidly used up; with 90 per cent. of sand, the soil becomes dusty in dry weather, and it becomes difficult to reap any benefit from it. With less than 30 per cent. of sand, the very clayey soils are still fitted for the cultivation of oats. When the proportion of sand is 30 per cent. the barley raised is better than the wheat.

THE CHEMICAL PROPERTIES OF SOIL.

A knowledge of the chemical composition of a soil is often of great benefit to the farmer, as allowing him to judge whether it contains the proper soil-constituents of which the crop he proposes to raise stands in need, or, being deficient, what is likely to prove the best fertilizer to be applied. Mere analytical results do not, in a great many cases, show the agricultural capabilities of a soil; thus, there are many soils whose chemical composition is apparently similar, that is to say, that the numerical results obtained by analysis show the like quantities of silica, lime, magnesia, soda, potash, phosphoric acid, &c., and yet a certain crop, clover for instance, will flourish on the one and not on the other. The physical nature of such soils, their depth, character of subsoil, aspect, texture, climatic conditions, &c., have likewise to be taken into account. Thus the many problems that enter into the study of soils are so various that chemical analysis alone does not afford, in most cases, a sufficient guide to an estimate of their agricultural capabilities, nor to point out the particular manure that is adapted for the special crops intended to be grown.

The most detailed chemical analysis usually gives only the proportion of the different constituents, and without any reference to their state of combination in which they exist in the soil or to their absorptive and retentive powers.

GREAT CARE NECESSARY IN OBTAINING THE SAMPLE FOR ANALYSIS.

On the care with which the soil is sampled, of course, depends the value of the analytical results, and too much stress cannot be laid on the necessity that exists to obtain a fair average sample, representing, as far as possible, both the good and bad qualities of the soil that is to be submitted to analysis. As the chemical analysis of a soil is a very long, tedious, and delicate operation, and the difference of a one-thousandth of 1 per cent. in any one constituent is equivalent to 20 or 30 pounds to the acre lost or gained in that element, the importance of the sample truly representing the soil is apparent.

THE CHEMICAL COMPOSITION OF SOILS.

Soil consists of an *organic* and of an *inorganic* or mineral part; the former derived from the decay of plant life for many ages, together with the dung and remains of animals, and the latter arising from the weathering of the rocks.

THE ORGANIC MATTER.

This varies in different soils, being most deficient in sandy soils and poor clays, and even in very fertile lands occurring only in small quantities. In the famous black soil of Russia, which is found in the provinces of the Ural Mountains and in those that border them, it varies from 5 to 12 per cent. In some of our own prairie soils the amount is nearly as high. In leaf mold it occurs considerably higher and in peat more than 50 per cent. very often. From its dark color it is a good absorbent of heat, its own specific heat being much above that of the soil generally. It is hygroscopic and greatly increases the water-holding power of sandy soil; besides, it has the power of absorbing and retaining ammoniacal salts. By its decomposition it forms a source of carbonic acid, which is readily absorbed by plant life. The mechanical condition of a soil is much improved by its presence when in moderate quantities; but when present in excessive amount, it acts injuriously by deoxidizing ferric salts and in other ways.*

THE INORGANIC OR MINERAL PORTION OF THE SOIL

Is, with the addition of alumina, composed of the same substances as make up the inorganic portion of plants, and which form their ashes when burnt. The mineral soil constituents include the following substances: Silica, SiO_2; alumina, Al_2O_3; lime carbonate, $CaCO_3$; ferric oxide, Fe_2O_3; phosphoric acid, P_2O_5 (phosphoric anhydride); potash, K_2O; Soda, Na_2O; magnesia, MgO; chlorine, Cl; sulphuric acid, SO_3 (sulphuric anhydride).

These exist in very different proportions in different soils. The first three, sand, clay, and lime, represent more than 90 per cent. of the substance of most soils, and as one or the other predominates the soil is said to be sandy, clayey, or calcareous. The most active constituents of the soil, phosphoric acid and the two alkalies, potash and soda, occur in very small quantities, as do the other and less important constituents, magnesia, chlorine, and sulphuric acid.

Silica exists in different proportions in the various soils, mostly in an insoluble state, and that most largely in the poorest sandy soils. Fertile soils contain generally a very small quantity of it in a soluble form. Sandy soils contain from 70 to 90 per cent. of silica, even stiff clay soils from 60 to 70 per cent., and calcareous or lime soils and marls from 20 to 30 per cent.

Its value, as a source of plant food, consists in being in the form of soluble silicates. In its insoluble state, like quartz sand, its action is merely mechanical, making the soil lighter for cultivation. Those soils, derived from rocks of which feldspar is one of the constituents, will contain some silica in a soluble form, whilst those derived from quartzose rocks will contain it in the insoluble state. The hydrated silica, in the analyses, represents that which is gradually avaliable for plant food.

Alumina or clay is a silicate of aluminium, and is derived from the disintegration of feldspathic rocks and other similar silicates; if absolutely pure it would furnish nothing for plant food; as, however, this is seldom the case, it furnishes a supply of potash frequently in considerable quantities. Clay has the important property of absorbing and retaining phosphoric acid, ammonia, potash, lime, and other substances necessary for plant food. Clay soils contain on an average from 6 to 10 per cent. of alumina. In sandy soils it varies from 1 to 4 per cent.; and in marls, calcareous soils, and vegetable molds from 1 to 6 per cent.

* Versuchs Stationen Organ, vol. xiv, p. 248–300.

The presence of alumina in the soil is purely mechanical, as it is never found in the mineral portions of plants, and the larger the percentage of it present the more difficult the soil becomes to cultivate, offering a greater or less resistance to the implements of tillage.

The percentage of alumina, as found by the method of chemical analysis used, is but an imperfect indication of the amount of *clay* in the soil. The amount of alumina continues to increase long after the rest of the important substances have been dissolved, if the digestion in hot dilute acid be prolonged. If this was combined as a hydrous silicate the amount of hydrated silica found, by boiling the insoluble residue with sodium carbonate, should bear a certain ratio to the alumina present. This, however, is seldom the case.

" It is but rarely that the amount of silica dissolved satisfies the requirement for combining with the alumina into kaolinite, and in a *very* few cases there is an excess of silica over that requirement. In numerous cases the silica falls so far below the amount corresponding to the alumina as to raise a serious question as to the combination in which the latter occurs in the soil, the *hydrate* (Gibbsite) being almost the only possible one, apart from zeolitic minerals. Perhaps this fact may serve to explain some of the otherwise incomprehensible variations in the physical properties of soils whose chemical and mechanical analysis would seem to make them almost identical. In some of the Tertiary and prairie soils of the Southern States, moreover, there seems to occur still another amorphous mineral, related to or identical with *Saponite*, which sometimes occurs in segregated masses, and imparts to these soils very peculiar and unwelcome properties in tillage. We are evidently, as yet, very far from a full understanding of the mechanical constitution of soils." (E. W. Hilgard, Tenth Census U. S.)

The lime or calcareous matter, generally occurring in the state of carbonate, varies in soils from about 90 per cent. and under in limestones and marls, to mere traces in some other soils. Clays and loams generally contain from 1 to 3 per cent. of the carbonate. Less than 1 per cent· may be regarded as a defective quantity. In the lightest sandy soils the percentage of lime should not fall below 0.100, in clay loams not below 0.250, and in heavy clay soils not below 0.500. (Hilgard.) Where a soil is deficient in lime, the little there is of it is present in combination with the organic acids, and is more abundant on the surface than in the subsoil. It preserves the particles of clay in a seperate coagulated condition, and thus allows them to exercise their absorbent powers on various salts, which otherwise would escape their action. It also promotes the decomposition of vegetable matter and the formation of nitrates in the soil.

Most green crops are often subject to disease, when grown on soils deficient in lime, even when they have been well manured. Up to a certain stage the cereal or other crops appear to thrive well, but as the season advances they sustain a check, and yield a poor harvest. This is especially the case in poor sandy soils, and a good dose of lime or marl, followed by barn-yard manure or guano has a most beneficial effect. By this means the valuable portion of the manure or guano, the ammonia, potash, and phosphoric acid, are retained in the land, whilst the others combine with the lime and are gradually washed out.

Ferric oxide is found in all soils, and causes the reddish color so very common in a great many of them. To its presence is chiefly due the retention of the phosphoric acid, an insoluble basic phosphate of iron being produced. On its state of oxidation depends its favorable influence on the soil, that of ferric, sesqui, or per oxide, better known as the red rust of iron, being the most suitable. In its less perfectly oxidized forms, which are, however, soluble in organic acids, it exists very often in the subsoil, and becomes peroxidized on exposure to the air. Its action is both physical and chemical. The preference of farmers for "red lands" arises from their experience of its beneficial action in the soil.

From 1.5 to 4 per cent. of ferric oxide is ordinarily found in soils but slightly tinted. Ordinary ferruginous loams vary from 3.5 to 7 per cent., highly colored "red lands" have from 7 to 12 per cent., and occasionally 20 per cent. and more. The efficiency of the ferric oxide depends upon its mechanical condition; when encrusting the grains of sand, or occurring as nodules, whilst the chemical analysis may show a large percentage of it present, it exerts little or no influence upon the soil, but when in a state of fine division these advantages are realized.

Soils containing a large percentage of ferric oxide have generally a low percentage of organic matter, but, notwithstanding, are as a rule very fertile. In clay lands especially its presence is very beneficial as tending to made them easier for tillage; its color tends to the absorption of heat and of oxygen. Such soils, however, suffer from floods or bad drainage, the ferric oxide becoming reduced under such circumstances to the ferrous state. (Hilgard.)

Phosphoric acid is contained in all good soils, but in very small quantities when compared with the other principal ingredients, and exists in combination with lime, iron, and alumina, phosphate of lime being its most common form. In general, even in the most fertile soils, it is found in very minute quantities, on an average less than a half per cent.; in clay lands this may rise to 1 per cent. Its value in fertilizers depends on its state of combination, whether it is soluble and immediately available for plant food, as the superphosphates, or slowly soluble, like the lime phosphates, forming a reserve store of food for the future. It occurs in all soils that have been formed from such rocks as the granites, gneisses, limestones, and dolomites, which contain it without exception; volcanic soils possess it in large quantity, whilst alluvial soils and those lands that are periodically swept by floods are much poorer. Soils containing less than 0.05 per cent. of it will be sterile and unfertile, as a general rule, unless accompanied by a large amount of lime.

Potash.—All soils suitable for cultivation contain potash in an available form, arising from the disintegration of feldspathic and other rocks. In the majority of cases the natural supply of the soil is sufficient to furnish to the plants the potash of which they are in need; a soil containing 0.125 per cent. should furnish potash enough for a century, without it being necessary to add to the manures used on such soils any salts of potash. Besides this available potash the soil often contains very considerable quantities of this element which the acids do not attack, and which form the reserve for the future supply of the plants. (Peligot.)

The quantity of potash varies in the different soils from the merest traces up to 1 and 2 per cent. Sandy and peaty soils and marls are generally deficient in this alkali, whilst soils rich in alumina are, with some exceptions, also rich in potash. It exists in the soil in combination with silica, forming a silicate, which is somewhat soluble in water. Heavy clay soils and clayey loams vary from 0.8 to 0.5 per cent; lighter loams from 0.45 to 0.30 per cent.; sandy loams below 0.3 per cent.; and sandy soils of great depth may contain less than 0.1 per cent. consistently with fertility, depending on the amounts of lime and phosphoric acid with which it is associated. (Hilgard.) A high percentage of potash in a soil seems capable of making up for a low percentage of lime, and conversely, a soil very rich in lime and phosphoric acid may be very fertile notwithstanding a low percentage of potash. The average annual consumption of potash for raising crops is 45 pounds per acre, or about 0.002 per cent.

Soda.—This is a less important constituent in soil than potash, and unless near the sea-coast, is present in even smaller quantities. Under the form of common salt, however, its presence is a cause of sterility in the soil when this exceeds 0.10 per cent. in quantity.

Magnesia is found in all fertile soils, in different proportions, often amounting to a mere trace. In the majority of cases the percentage of magnesia is greater than that of the lime, but it does not seem capable of performing to any appreciable extent the general function of lime in soil improvement.

Sulphuric acid and chlorine occur very sparingly in most soils. From 0.02 to 0.04 per cent. of the former seems to be adequate to most soils.

There does not exist any affinity between the quantities of lime and magnesia contained in soils and those of potash and phosphoric acid.

Nitrogen and nitrates.—The natural sources of nitrogen in crops are the nitrates and ammonia salts, which are seldom present in large quantities, and should be used on or generated in the soil as rapidly as crops require them. The process of nitrification, whereby inert or unassimilable nitrogen becomes converted into nitric acid, is thus of great importance to agriculturists. This is due to a minute *Bacterium*, present in all soils, whereby the humus and ammonia are oxidized and their nitrogen converted into nitric acid. This process does not take place unless the soil is moist and has free access of air, and some base, generally lime, is present with which the nitric acid can combine. Nitrification is thus most active in summer and ceases apparently in winter.

Messrs. Lawes and Gilbert have for some years past been devoting their attention to the sources of the nitrogen of crops, and in the pages of the Journal of the Royal Agricultural Society, and of the Journal of the Chemical Society, will be found their reports in full.

The following is the summary and conclusion which they have just published in a long article on "Some points in the composition of soil," in the June number of the Chemical Society for last year, p. 420:

(1) The annual yield of nitrogen per acre in various crops, grown for many years in succession on the same land without nitrogenous manure, was found to be very much greater than the amount of combined nitrogen annually coming down in rain and the minor measurable aqueous deposits.

(2) So far as the evidence at command enables us to judge, other supplies of combined nitrogen from the atmosphere, either to the soil or to the plant itself, are quite inadequate to make up the deficiency.

(3) The experimental evidence as to whether plants assimilate the free nitrogen of the atmosphere is very conflicting; but the balance is decidedly against the supposition that they so derive any portion of their nitrogen.

(4) When crops are grown year after year on the same land, for many years in succession without nitrogenous manure, both the amount of produce per acre, and the amount of nitrogen in it, decline in a very marked degree. This is the case even when a full mineral manure is applied; and it is the case not only with cereals and with root crops, but also with *Leguminosæ*.

(5) Determinations of nitrogen in the soils show that, coincidentally with the decline in the annual yield of nitrogen per acre of these very various descriptions of plants, grown without nitrogenous manure, there is also a decline in the stock of nitrogen in the soil. Thus a soil source, of at any rate some, of the nitrogen of the crops is indicated. Other evidence pointed in the same direction.

(6) Determinations of the nitrogen as nitric acid, in soils of known history as to manuring and cropping, and to a considerable depth, showed that the amount of nitrogen in the soil in that form was much less after the growth of a crop than under corresponding conditions without a crop. This was the case not only with gramineous but with leguminous crops. It was hence concluded that nitrogen had been taken up as nitric acid by the growing crops.

(7) In the case of gramineous crop soils, the evidence pointed to the conclusion that most, if not the whole, of the nitrogen of the crops was taken up as nitric acid from the soil.

(8) In the experiments with leguminous crop soils, it was clear that some at any rate of the nitrogen had been taken up as nitric acid. In some cases, the evidence was in favor of the supposition that the whole of the nitrogen had been so taken up. In others this seemed doubtful.

(9) Although in the growth of leguminous crops year after year on the same land without nitrogenous manure, the crop, the yield of the nitrogen in it, and the total nitrogen in the surface soil, greatly decline, yet, on the substitution of another plant of the same family, with different root habits and root range, large crops, containing large amounts of nitrogen, may be grown. Further, in the case of the occasional growth of a leguminous crop, red clover for example, after a number of cereal and other crops, manured in the ordinary way, not only may there be a very large amount of nitrogen in the crop, presumably derived from the subsoil, but the surface soil becomes determinably richer in nitrogen, due to crop residue.

(10) It was found that, under otherwise parallel conditions, there was very much more nitrogen as nitric acid, in soils and subsoils down to a depth of 108 inches, where

leguminuous than where gramineous crops had grown. The results pointed to the conclusion that, under the influence of leguminous growth and crop residue, the conditions were more favorable for the development of the nitrifying organisms, and, especially in the case of deep-rooting plants, of their distribution, thus favoring the nitrification of the nitrogen of the subsoil, which so becomes a source of the nitrogen of such crops.

(11) An alternative was that the plants might take up at any rate part of their nitrogen from the soil and subsoil as organic nitrogen. Direct experimental evidence leads to the conclusion that fungi take up both organic nitrogen and organic carbon, but there is at present no direct experimental evidence in favor of the view that green-leaved plants take up either nitrogen or carbon in that form from the soil; whilst there are physiological considerations which seem to militate against such a view.

(12) In the case of plots where *Trifolium repens* (white clover) and *Vicia sativa* (tares or vetches) had been sown, each for several years in succession, on soil to which no nitrogenous manure had been applied for about thirty years, and the surface soil had become very poor in nitrogen, both the soil and subsoil contained much less nitrogen as nitric acid where good crops of *Vicia sativa* had grown, than where the more shallow-rooted *Trifolium repens* had failed to grow ; and the deficiency of nitric nitrogen in the soils and subsoils of the *Vicia sativa* plots, compared with the amount in those of the *Trifolium repens* plot, was, to the depth examined, sufficient to account for a large proportion of the nitrogen of the *Vicia* crops.

(13) It may be considered established, that much, if not the whole, of the nitrogen of crops is derived from nitrogen within the soil, accumulated or supplied; and that much, and in some cases the whole, of the nitrogen so derived, is taken up as nitrates.

(14) An examination of a number of United States and Canadian prairie soils showed them to be very much richer in both nitrogen and carbon, to a considerable depth, than the surface soils of old arable lands in Great Britain, and about as rich, to a much greater depth, as the surface soil of permanent pasture land.

(15) On exposure of portions of some of these rich prairie soils, under suitable conditions of temperature and moisture, for specified periods, it was found that their nitrogen was readily susceptible of nitrification, and so of becoming easily available to vegetation.

(16) After several extractions, the subsoils almost ceased to give up nitric acid ; but on seeding them with a tenth of a gram of rich garden soil containing nitrifying organisms, there was a marked increase in the rate of nitrification. This result afforded confirmation of the view that the nitrogen of subsoils is subject to nitrification, if under suitable conditions, and that the growth of deep-rooted plants may favor nitrification in the lower layers.

(17) Under favorable conditions of season and of cultivation the rich prairie soils yield large crops; but, under the existing conditions of early settlement, they do not, on the average, yield crops at all commensurate with their richness, when compared with the soils of Great Britain which have been under arable culture for centuries. But so long as the land is cheap, and labor dear, some sacrifice of fertility is unavoidable in the process of bringing these rich virgin soils under profitable cultivation.

(18) A comparison of the percentages of nitrogen and carbon in various soils of known history, showed that a characteristic of a rich virgin soil, or of a permanent pasture surface soil, was a relatively high percentage of nitrogen and carbon. On the other hand, soils which have long been under arable culture are much poorer in these respects; whilst arable soils under conditions of known agricultural exhaustion, show a very low percentage of nitrogen and carbon, and a low proportion of carbon to nitrogen.

(19) Not only the facts adduced in this and former papers, but the history of agriculture throughout the world, so far as it is known, clearly shows that, pre-eminently so far as the nitrogen is concerned, a fertile soil is one which has accumulated within it the residue of ages of natural vegetation, and that it becomes infertile as this residue is exhausted.

The following table shows the character of exhausted arable soils, of newly laid down pasture lands, and of old pasture soils at Rothamsted, England; of some other old arable soils; of some Illinois and Manitoba prairie soils; and, lastly, of some very rich Russian soils, in regard to their percentages of nitrogen and carbon, taken from the same report, p. 419:

TABLE II.—*Nitrogen and carbon in various soils.*

.	Date of soil sampling.	In dry-sifted soil.*			Authority.
		Nitrogen.	Carbon.	Carbon to 1 nitrogen.	
Rothamsted arable and grass soils.					
		Per cent.	Per cent.	Per cent.	
Roots, 1843-'52; Barley, 1853-'55; roots, 1856-'69; mineral manures.	Apr., 1870	0.0934	....	....	Rothamsted.
Wheat, 1843-'44, and each year since; mineral manures.	Oct., 1865	0.1119	1.030	9.3	Do.
	Oct., 1881	0.1012	1.079	10.7	
Barley, 1852, and each year since; mineral manures.	Mar., 1868	0.1202	......	......	Do.
	Mar., 1882	0.1124	1.154	10.3	
Arable laid down to grass (Fen acres), spring, 1879.	Feb., 1882	0.1235	....	....	Do.
Arable laid down to grass (Barn-field), spring, 1874.	Feb., 1882	0.1509	....	....	Do.
Arable laid down to grass (Apple-tree field), spring, 1863.	Nov., 1881	0.1740	....	....	Do.
Arable laid down to grass (Dr. Gilbert's meadow), spring, 1858.	Jan., 1879	0.2057	2.412	11.7	Do.
Arable laid down to grass (High-field), spring (?), 1838.	Sept., 1878	0.1943	2.403	12.4	Do.
Very old grass land (the Park)	Feb., 1876 Mar., 1876	0.2466	3.377	13.7	Do.
Various arable soils in Great Britain.					
Mr. Prout's farm; Broadfield, surface	....	0.170	....	....	Voelcker.
Mr. Prout's farm; Blackacre, surface	....	0.107	....	....	Do.
Mr. Prout's farm; Whitemoor, surface	....	0.171	....	....	Do.
Wheat soil:					
Midlothian	....	0.220	....	....	Anderson.
Eastlothian	....	0.130	....	....	Do.
Perthshire	....	0.210	....	....	Do.
Berwickshire	....	0.140	....	....	Do.
Red sandstone soil, England	....	0.180	....	....	Voelcker.
United States and Canadian prairie soils.					
Illinois, United States, No. 1	....	0.300	....	....	Do.
Illinois, United States, No. 2	....	0.260	....	....	Do.
Illinois, United States, No. 3	....	0.330	....	....	Do.
Illinois, United States, No. 4	....	0.340	....	....	Do.
Portage la Prairie, Manitoba, surface	....	0.247	....	....	Rothamsted.
Saskatchewan district, Northwest Territory, surface.	....	0.303	....	....	Do.
Forty miles from Fort Ellis, Northwest Territory, surface.	....	0.250	....	....	Do.
Niverville, Manitoba, first 12 inches	....	0.261	3.42	13.1	Do.
Brandon, Manitoba, first 12 inches	....	0.187	2.66	14.2	Do.
Selkirk, Manitoba, first 12 inches	....	0.618	7.58	12.3	Do.
Winnipeg, Manitoba, first 12 inches	....	0.428	5.21	12.2	Do.
Russian soils.					
No. 1. 12 inches	....	0.607	....	....	C. Schmidt.
No. 2. 8 inches	....	0.467	....	....	Do.
No. 3. 5 inches	....	0.188	....	....	Do.
No. 4. 6 inches	....	0.130	....	....	Do.
No. 5. 11 inches	....	0.305	....	....	Do.
No. 6. 17 inches	....	0.281	....	....	Do.
No. 7. 9 inches	....	0.409	....	....	Do.

* Calculated on soil dried at 100° C.

FERTILITY DEPENDS ON THE MINIMUM QUANTITY OF ANY NECES-SARY CONSTITUENT PRESENT.

As the soil is the source whence plants derive their mineral food, all the elements required for this nourishment have, in a certain sense, equal value; for if one of them is wanting in the soil, or is present in a form not readily available by the roots, the plant will not flourish; and so its fertility depends on the *minimum* quantity of any necessary constituent present.

WEIGHT OF A SOIL PER ACRE.

The weight of soil on an acre of land is so enormous that even small proportions of plant food may amount to very considerable quantities. The weight varies with the porosity and the amount of sand and gravel the soil contains.

The following table gives the specific gravity, the weight of 1 cubic foot of different soils, dry and wet, taken from the table in Professor Schübler's article "On the Physical Properties of Soil," in the Journal of the Royal Agricultural Society, vol. 1, p. 210; together with their approximate weight per acre to a depth of 9 inches, equal to 32,670 cubic feet, in tons of 2,000 pounds:

TABLE III.—*Table showing the specific gravity, the weight of one cubic foot of different soils, dry and wet, according Schübler, and the approximate weight per acre to a depth of 9 inches (32,670 cubic feet).*

Kind of soil.	Specific gravity.	Weight one cubic foot.		Weight per acre, 9 inches deep.	
		Dry.	Wet.	Dry.	Wet.
		Pounds.	*Pounds.*	*Tons.*	*Tons.*
Siliceous sand, occurring in almost every arable soil	2.653	111.3	136.1	1,818.0	2,223.2
Calcareous sand, frequently occurring along with siliceous sand	2.722	113.6	141.3	1,855.6	2,308.1
Sandy clay, a combination of 45 per cent. of fine sand with 55 per cent. of clay	2.601	97.8	129.7	1,597.6	2,118.6
Loamy clay, a combination of 24 per cent. of fine sand with 76 per cent. of clay	2.581	88.5	124.1	1,445.6	2,027.2
Stiff clay, a combination of 10 per cent. of fine sand with 90 per cent. of clay	2.560	80.3	119.6	1,311.7	1,953.7
Clay, in its fine pure state, 58 per cent. of silica, 36.2 per cent. of alumina with 5.8 per cent. of ferric oxide	2.533	75.2	115.8	1,228.4	1,891.6
Slaty marl	2.631	112.0	140.3	1,829.5	2,291.8
Humus	1.370	34.8	81.7	568.5	1,334.6
Fertile garden mold	2.332	68.7	102.7	1,122.2	1,677.6
Common arable soil	2.401	84.5	119.1	1,380.3	1,945.5

Thus 0.10 per cent. of any constituent, such as phosphoric acid, potash, &c., would amount to from 2,250 to 3,500 pounds in 1 acre of soil 9 inches deep.

QUESTIONS OFTEN ANSWERED BY THE ANALYSIS OF SOILS.

The results of soil analysis frequently give decided and satisfactory answers, according to Dr. Voelcker,* to the following questions:

(1) Whether or not barrenness is caused by the presence of an injurious substance, such as sulphate of iron or sulphide of iron, occurring in peaty and clayey soils?

(2) Whether soils contain common salt, lands flooded by sea water, nitrates or other soluble salts, that are useful to vegetation in a highly diluted state, but injurious when they occur in land too abundantly?

(3) Whether barrenness is caused by the absence or deficiency of lime, phosphoric acid, or other important elements of plant food?

(4) Whether clays are absolutely barren, and not likely to be materially improved by cultivation, or whether they contain the necessary elements of fertility in an un_available state, and are capable of being rendered fertile by subsoiling, deep cultivation, steam plowing, and similar mechanical means?

(5) Whether or not clays are usefully burnt and used in that state as manure?

(6) Whether or not the land will be improved by liming?

(7) Whether it is better to apply lime, or marl, or clay, on a particular soil?

(8) Whether special manures, such as superphosphates or ammoniacal salts, can be used—of course, discreetly—without permanently injuring the land; or whether the farmer should rather depend upon the liberal application of farm-yard manure, that he may restore to the land all the elements of fertility removed in the crops?

(9) What kind of artificial manures are best suited to soils of various compositions?

According to the same authority,† chemical analysis cannot supply any definite information in regard to barrenness of soil on the following questions:

(1) Whether barrenness is caused by defective drainage?

(2) To what extent sterility is affected by a bad physical condition of the land?

(3) How far unproductiveness is affected by the climate?

(4) That a soil is barren simply because there is too little of it; or

(5) That it is unproductive simply because a thin surface soil rests on a stiff clay subsoil of great depth?

(6) What is the relative productiveness of different soils?

OBJECTS AND INTERPRETATION OF SOIL ANALYSIS.

For a very full discussion of the objects and interpretation of soil analysis the reader is referred to an article on this subject in the American Journal of Science vol. 22, pp. 183–197, by Prof. E. W. Hilgard, as well as the report on "Soil Investigation," by the same author, contained in the "General Discussion of the Cotton Production of the United States," Tenth Census of the United States, 1880, Vol. V, pp. 67–81, of which the following is a summary:

The claim of soil analysis to practical utility has always rested on the general supposition that, "other things being equal, productiveness is, or should be, sensibly proportional to the amount of available plant food within reach of the roots during the period of the plant's development;" provided, of course, that such supply does not exceed the maximum of that which the plant can utilize when the surplus simply remains inert. This statement is, either tacitly or expressly, admitted by all those

---

* Journ. Royal Agric. Soc., vol. xiv, p. 338.
† Journ. Royal Agric. Soc., vol. 1, 1865, p. 129.

who have attempted to interpret soil analyses, and agrees with the accumulated experiences of agriculturists.

Many attempts have been made to find solvents that shall represent correctly the action of the plant itself on the soil ingredients, in order that conclusions might be made as to the present agricultural value of a given soil. From sulphuric and hydrofluoric acids to water charged with carbonic acid, as used by Dr. D. D. Owen, the acid solvents have all signally failed to secure even an approximation to the result desired, viz, a consistent agreement between the quantitative determination of the plant food found in the several soils, and the actual experience of those who cultivate them.

The ultimate analysis of soils, as attempted by the German experiment stations, under Wolff's initiative, by the consecutive extractions with acid solvents of different strengths, beginning with distilled water, and ending with boiling sulphuric or hydrofluoric acids, affords little or no clew to their agricultural value.

Soil analyses do not, like the assay of an ore, interpret themselves to a layman; a column of figures summing up to 100 or nearly so, opposite another column of unintelligible names does not convey much information to a farmer.

In Europe, and in the thickly settled portions of this country, the arable soils have nearly all been at some time subjected to cultivation, and to the use of fertilizers, thus veiling their original characteristics and rendering extremely difficult the taking of any sample of soil that shall represent correctly, in all respects, the whole of any large field or district. In the greater portion of this country, however, we are able to procure samples of the virgin soil that even the plow has not touched nor any manures been applied. The virgin soil and its vegetation are the outcome of long ages of coadaption by the process of natural selection; and the settler is afforded a means of judging of the productiveness and durability of the land based upon the character of its vegetation.

A soil naturally timbered with a large proportion of walnut, wild cherry, or, as at the South, with the "poplar" or tulip tree, is at once selected as sure to be both productive and durable, especially if the trees be large. The settler knows well that the black and Spanish oaks frequent only "strong soils," and an admixture of hickory is a welcome addition; while the occurence of the scarlet oak at once lowers the land in his estimation, and that of pine still more so.

Having obtained the percentage composition of a soil, how are we to interpret these percentages to the farmer? What are "high" and "low" percentages of each ingredient important to the plant, whether as food or through its physical properties?

The first question is, naturally, whether all soils, having what experience proves to be high percentages of plant food when analyzed by the method given elsewhere, show a high degree of productiveness. This question can be unqualifiedly answered in the affirmative in regard to virgin soils, provided only that improper physical conditions do not interfere with the welfare of the plant. But it does not follow that the converse is true, and that low percentage necessarily indicates low production.

For instance, we may have a heavy alluvial soil of high percentages and producing a maximum crop in favorable seasons. If this soil be mixed with its own weight, or even more, of coarse sand, thereby reducing the percentages one-half or less; and yet it will not only not produce a smaller crop, but it is more likely to produce the maximum crop every year, on account of improved physical conditions. If we compare the root system of the plants grown in the original and in the diluted soil, we will find the roots in the latter more fully diffused, longer, and better developed, not confined to the crevices of a hard clay, but permeating the entire mass, and evidently having fully as extensive a surface contact with the fertile soil particles as was the case in the original soil. How far this dilution may be carried without detriment would vary with different plants and soils, and must largely be a matter of experiment. A plant capable of developing a very large root surface can obviously make up by greater spread for a far greater dilution than one whose root surface is in any case but small. The former flourishes even on "poor, sandy" soils, whilst the latter succeeds, and is naturally found on "rich, heavy" ones only, although the absolute amount of plant food taken from the soil may be the same in either case.

It is obvious that without a knowledge of the respective depths and penetrability of two soils a comparison of the percentages of their plant constituents will be useless. The surface soil, with its processes of nitrification, oxidation, carbonic acid solution, &c., in full progress, must always be distinguished from the subsoil in which these processes are but feebly developed, and where the store of plant food, in which it is generally richer than the surface soil, is comparatively inert. Hence the obvious importance of samples correctly taken, and the necessity of intelligent and accurate observations on the spot.

The concentration of the available portion of the plant food of soils in their finest portions has become a maxim. A "strong soil" is invariably one containing within reach of the plant the large amount of impalpable matter; although the reverse is by no means generally true.

A comparison of the composition of soils of known productiveness, and characterized in their natural state by certain invariable features of plant growth, soon reveals the existence of definite relations, not only to the *absolute amounts* of certain ingredients present in the soil, but also to their *relative proportions*. No ingredient exerts in these respects a more decided influence than *lime*, its advent in relatively large proportion, other things remaining equal, changing at once the whole character of vegetation, so as to be a matter of popular remark everywhere. Only it is not popularly known, nor has it been definitely recognized by agricultural chemists thus far, that it is the *lime* that brings the change.

The amount of the different soil constituents which may be considered the minimum consistent with fertility has already been given.

THE METHOD OF SOIL ANALYSIS.

The following method for the analysis of soils has been adopted by this division, and is essentially the same as that described by Sprengel and Otto in Sprengel's Bodenkunde, 1837, p. 370, depending on the principle that in order to judge of the fertility of a soil, it is necessary to determine not only what are its elementary constituents, but likewise the manner in which they are combined. With this view Sprengel and Otto treated a sample of the soil successively with water, dilute hydrochloric acid, strong sulphuric acid, and by fusion with an alkaline carbonate. The constituents of the soil soluble in water were supposed to be actually available as plant food; those which were soluble in dilute acids as available for that purpose after being subjected to the action of carbonic acid and the humous acids of the soil; the rest as available only after the soil had been subjected for a considerable time to atmospheric influences, assisted by the mechanical operations of tillage. This mode of proceeding was originally based upon the supposition that the constituents of the soil which are to nourish the plant must be presented to it in the state of solution, a view which can no longer be regarded as correct. Hence the treatment of the sample of soil with water, with strong sulphuric acid, and by fusion with an alkaline carbonate, has not been followed, because it was considered that the estimation of the constituents soluble in dilute acids was the most important as showing the total amount of those constituents present in the soil as readily available as plant food. This was formerly made on the residue left after exhausting the soil with boiling water, but, as this mode of operation is no longer regarded as of much importance, the soil is now directly treated with the dilute acids after the removal of the stones and coarser particles.

The mode of operation pursued and the precautions to be observed have been stated at length, as where such small quantities are obtained the greatest care is necessary in all manipulations to obtain accurate results. References have been made, in all cases, to Fresenius's Quantitive Analysis. Results obtained from different portions of the same sample of soil have been found to agree very closely, thus establishing the accuracy of the methods employed, where all precautions are observed.

The strength of the acids used and the time employed for digestion is that found by Professor Hilgard, from investigations made by Dr. R.

H. Loughbridge,* to exert the maximum effect after a water-bath digestion of five days.

In determining the different soil constituents especial care must be taken to subject them in the various operations to as nearly similar conditions as possible, as the principal object is to obtain comparative results; to this end the several soils are air-dried together, digested for an equal time at nearly the same temperature in acids of uniform strength, &c.

Owing, however, to the large amount of time demanded by such work, the process of digestion in water containing carbonic acid was not employed, neither was any attempt made at a mechanical analysis of the soils received.

COLLECTING THE SAMPLE.

The collection of samples of soil is a delicate and important operation, as it is on the *average sample* that the physical and chemical properties of the soil are determined. They should represent, as far as possible, the average of the bad and good qualities of the soil.

Select, in the field, four or five places, at least, per acre, taking care that these places have an homogeneous aspect, and represent as far as possible the general character of the whole ground. If the field, however, presents notable differences, either in regard to its aspect or its fertility, the sample gathered from the different parts must be kept separate.

The sampling of the arable soil should be made only after the raising of the crop and before it has received any new manure, in the following manner:†

Have a wooden box made, 6 inches long and wide, and from 9 to 12 inches deep, according to the depth of soil and subsoil in the field. At one of the selected places mark out a space of 12 inches square; dig around it in a slanting direction a trench, so as to leave undisturbed a block of soil, with its subsoil, from 9 to 12 inches deep; trim this block to make it fit into the wooden box, invert the open box over it, press down firmly, then pass a spade under the box and lift it up and gently turn over the box.

In the case of very light, sandy, and porous soils, the wooden box may be at once inverted over the soil and forced down by pressure, and then dug out.

Proceed in the same way for collecting the samples from all the selected places in the field, taking care that the subsoil is not mixed with the surface soil. The former should be sampled separately.

In preparing the plot for the gathering of the sample, take care to have it lightly scraped so as to remove any débris which may be accidentally found there.

The sample should be taken only from spots that have not been cultivated, where "virgin soils" are concerned, and as a rule not from ground frequently trodden on, footpaths, roads, &c., nor from squirrel holes,

---

* Amer. Jour. of Science, vol. vii, p. 20.

† From the instructions for selecting samples, issued by the Royal Agricultural Society of England.

stumps, and the foot of trees, nor from spots washed by streams or rain, and are thus not a fair representative of the land. Avoid spots showing unusual growths, whether in kind or quality. Note carefully the normal vegetation, trees, herbs, grass, &c., the general character of the land, whether hilly, rolling, flat, &c., the aspect, elevation, and such peculiarities of the soil and subsoil, their behavior in wet and dry weather, the character of the crops raised on the land, in fact, every circumstance that can throw any light on their agricultural qualities or peculiarities. Unless accompanied by such notes and memoranda samples of soils cannot be considered as justifying the amount of time and labor involved in their chemical examination.

The "soil" is that portion of the surface of the ground which is reached by ordinary tillage operations, generally being from 6 to 9 inches deep; the "subsoil" is that portion which is ordinarily not touched in plowing, lying beneath the soil.

It is always well to know what constitutes the nature of the foundation of the soil to a depth of, at least, 36 inches, since the question of drainage, resistance to drought, &c., will be influenced by the character of the substratum.

The different samples thus procured are emptied on a clean, boarded surface, and thoroughly mixed, so as to incorporate the different samples of the same field together. The heap is then divided into four divisions, and the opposite quarters are put one side, taking care to leave the two remaining ones undisturbed; these are thoroughly mixed together, the heap divided into quarters, and the opposite ones taken away as before. This operation of mixing, dividing into quarters and taking away the opposite quarters, is continued until a sample is left weighing about 10 or 12 pounds.

Thus is obtained the *average sample* of the soil. Of course where only a single sample is taken from the field this method of quartering is not resorted to, but the bottom of the box is nailed directly on and sent to the laboratory, where the soil is to be analyzed.

PREPARATION OF THE SAMPLE.

The sample of the soil to be analyzed, after it is received in the laboratory and given an index number, is immediately spread out in a thin layer on a large shallow pan and fully exposed to the warm air of the room, until it is thoroughly dried at the common temperature, or, better, in an air-bath at a temperature of 60° to 100° C. When it is dry and friable it is carefully sampled, by the method of quartering, until a sample of about 150 grams is left. This is then rubbed up in a porcelain mortar, taking care to avoid grinding up any of the gravel or fragments of rocks, &c., which it may contain, and which are removed and estimated, and the fine powder is then passed through a wire sieve of 25 meshes to the square inch. The object of this operation is to bring the whole sample of the soil to a state of uniform mixture, and to remove from it the coarser gravel and roots, &c., which it may contain.

DETERMINATION OF MOISTURE AND OF VOLATILE AND ORGANIC MATTER.

Moisture.—Introduce from 2 to 5 grams of the air-dried soil into a previously weighed platinum dish holding about 10 c. c. and dry at 120° C. in an air-bath for eight hours, cool in a desiccator, and weigh. Repeat the heating and weighing until the substance ceases to lose weight or begins to increase, indicating incipient oxidation. From the lowest weight thus obtained calculate the percentage of moisture. The difference between the first weight of the platinum dish and substance and that found on drying represents the moisture, and this weight divided by the quantity taken and the quotient obtained multiplied by 100 will give the percentage. The results are only approximate, as the complete drying of a soil, especially if it contains much clay or organic matter, is very difficult to effect. Some soils are very hygroscopic and rapidly absorb the atmospheric moisture; for this reason the platinum dish should be cooled in a desiccator containing fused calcium chloride and rapidly weighed.

Volatile and organic matter.—The dried substance is then ignited at a low red heat, in a muffle furnace, until the whole of the organic matter has been destroyed, care being taken that the heat is not raised too high in order to avoid driving off any of the alkaline chlorides, &c.

The residue is ordinarily of a reddish color, owing to the sesquioxide of iron which it contains. When the mass is cool it is treated with a few drops, about 1 c. c., of a saturated solution of ammonium carbonate or oxalate, and then gradually heated to about 150° C. in the air-bath, avoiding all danger of loss by sputtering by a careful regulation of the heat at the commencement. By this means any carbonates that may have been decomposed by the ignition are reconverted.

The loss in weight represents the *organic* and other *volatile matters*.

TREATMENT OF THE SOIL WITH HYDROCHLORIC ACID (Sp. gr. 1.115).

In using the different reagents and distilled water especial care must be taken that they are all *chemically pure*, in order that no foreign matter may be introduced into the analysis by them; for this purpose they must be carefully tested. It is hardly necessary to add that all the weighing of precipitates, &c., must be done on a delicate analytical balance.

The hydrochloric acid used is made from the concentrated acid, C. P., by diluting with distilled water until it attains a specific gravity of 1.115 as shown by an hydrometer, or, better, taken in a specific gravity bottle, and its weight compared with that of an equal volume of water at the same temperature.

Ten grams of the air-dried soil are treated with 200 c. c. of hydro-chloric acid, specific gravity 1.115, for five days, on a steam-bath, at 100° C., in a large porcelain dish, covered with a clock-glass, keeping up the volume by the addition of distilled water when necessary. At the end of the five days the solution being, from time to time, stirred with a glass rod during this treatment, is allowed to cool and settle completely. Then filter off the solution and wash the residue of in-soluble matter well with boiling water, allowing it to settle before de-canting through the same filter. Repeat this washing by decantation two or three times, not using more than 50 c. c. of boiling water at a time. Finally transfer the insoluble matter to the filter, washing it thoroughly out of the dish with hot water. It often happens that some portions will remain sticking to the sides of the dish. These are removed by rub-bing them with a rubber-covered glass rod, and washing them onto the filter. A few drops of the wash-water are tested with argentic nitrate, on a watch-glass, to see that all the soluble chlorides have been washed out. This testing is not done until about 200 c. c. of water has passed through the funnel. If any chlorides be present in the wash-water a cloudiness is produced on the addition of the argentic nitrate solution, in which case the washing of the insoluble residue is con-tinued until, on testing, all the chlorides are removed. (Fres., §140, II, a.)

The filtrate A. and washings should not exceed 500°c. c.

DETERMINATION OF THE INSOLUBLE RESIDUE.

Dry the filter containing the insoluble residue A in the air-bath at 120° C. When dry transfer it to a previously weighed platinum cru-cible and ignite it carefully, at first, until the filter paper is consumed and then raise the heat until the crucible and contents are at a bright red and continue it until all the organic matter is consumed. The in-soluble residue should become white, or nearly so, by this treatment. Cool and weigh. The increase in weight gives that of the insoluble and hydrated silica, plus the filter-ash, which must be deducted in all cases.

DETERMINATION OF THE HYDRATED SILICA.

To ascertain how much of the silica found exists in combination with the bases of the clay, how much as hydrated acid, and how much as quartz sand, or as a silicate present in the form of sand, proceed as fol-lows: The insoluble residue A, after it has been ignited and weighed, is transferred, in small portions at a time, to a boiling, rather concen-trated, solution of sodium carbonate, contained in a large platinum dish holding about 200 c. c.; boil for some time, and filter off each time, still very hot. When all is transferred to the dish, boil repeatedly with the strong solution of sodium carbonate until a few drops of the fluid remain clear on warming with ammonium chloride.

Wash the residue, by decantation, several times with hot water and then transfer it to the filter, and to make sure of removing every trace of the sodium carbonate which may still adhere to it, with water slightly acidulated with hydrochloric acid, and finally with hot water.

This will dissolve the silica in combination with the bases of the clay, and also the hydrated silica, and leave a residue of quartz sand and silicates in the form of sand, e. g., feldspar sand, which is dried, ignited, and weighed.

The difference between this last weight and that of the insoluble residue A will give the amount of hydrated silica. Or this may be determined directly by acidifying the filtrate with hydrochloric acid, evaporating to dryness, and driving off the chlorine, taking up with dilute hydrochloric acid, filtering, washing well, drying, igniting, and weighing the hydrated silica. (Fres., § 209.).

DETERMINATION OF THE SOLUBLE SILICA.

Evaporate the main solution A to dryness in a porcelain dish, adding a little, about 2 c. c., of nitric acid to destroy the organic matter and oxidize the iron. Then heat in the air-bath at 110° C. until the acid fumes disappear. By holding a glass rod moistened with ammonia in the dish and noting when the white fumes cease to be produced, an easy means is afforded of telling when the hydrochloric acid is entirely driven off.

After the mass is thoroughly dried moisten it with 25 c. c. dilute hydrochloric acid, heat to a temperature just below boiling for twenty or thirty minutes, and dilute with 50 c. c. of hot water; everything should be in solution except the silica. Filter this out, and wash with hot water until the washings give no reaction for chlorine when tested with argentic nitrate. Dry the precipitate on the filter at 110° C., and transfer it to a weighed platinum crucible, and heat at a low temperature until the paper is consumed. Care must be taken that the heat is not too strong at first, as there is great danger of some of the silica passing off with the volatile matter. Gradually raise the heat until the silica becomes white. Then cool the crucible and contents in a desiccator, weigh, and calculate the per cent. of soluble silica. (Fres., § 140, II, a.)

DETERMINATION OF THE IRON AND ALUMINA.

Make the filtrate from the soluble silica up to 500 c. c., solution B.

To 200 c. c. of the solution, equal to 4 grams of the air-dried soil, add ammonia to alkaline reaction, to precipitate the aluminium and ferric hydrates, together with the phosphoric acid. Be careful *not* to add a large excess, or time will be wasted in boiling it out, which will be necessary, for the reason that aluminium hydrate is somewhat soluble in excess of ammonia. Boil the solution until the vapors no longer smell of ammonia, and do not turn turmeric paper brown. Allow the precipitate to settle and decant the clear fluid on a filter. Then wash with 50 c. c. of boiling water, stir, allow to settle, and decant as before. As some lime and magnesia may be carried down by the precipitation of the hydrates, dissolve it in the beaker, with as little dilute hydrochloric acid as possible, reprecipitate by adding some ammonia and boiling as before. Wash by decantation, three or four times, using 40 to 50 c. c. of boiling water each time. Finally transfer the precipitate to the filter, with boiling water, and wash with the same, until a few drops of the wash water, acidulated with nitric acid, do not show any trace of chlorides when tested with argentic nitrate.

Dry the filter and contents in an air-bath at 110° C., and when perfectly dry ignite it in a weighed platinum crucible, applying the heat gently at first, until the filter paper is consumed, and then more intensely, cool and weigh. This weight, after deducting that of the crucible and filter ash, will be that of the aluminium and ferric oxides with the phosphoric acid contained in the 4 grams of air-dried soil. (Fres. §105, a, §113, I, a.)

The weight of the phosphoric acid, determined by the method given further on, is deducted from this weight, thus leaving the weight of the two oxides.

DETERMINATION OF THE FERRIC OXIDE BY TITRATION WITH POTASSIUM PERMANGANATE.

The ignited and weighed precipitate of the aluminium and ferric oxides carrying the phosphoric acid is transferred to a small beaker, and the crucible carefully washed with water to remove any adhering particles. Concentrated sulphuric acid is then cautiously added, about 10 c. c. is sufficient, and digested on the steam-bath until all the substance is in complete solution. The solution is allowed to become cold, and is then diluted with about 150 c. c. of water, and is passed through a small filter into a 200 c. c. cylinder, and washed slightly. The solution is made up to the mark and divided into two equal portions of 100 c. c. each.

Each of these two portions is transferred to a reducing bottle. Place in each bottle a piece of amalgamated zinc, and a piece of platinum foil an inch wide and 4 or 5 inches long, add 2 c. c. concentrated sulphuric acid, fill with water to the shoulders, cover with watch-glasses and allow to stand twenty-four hours. A strong current of gas should be induced by contact between the zinc and the platinum. The zinc used must be amalgamated, as it usually contains iron, which, in dissolving, it will impart to the solution if this is not done. It has been found by experiment that amalgamated zinc will not give up the iron it contains to the solution until nearly if not quite all the zinc is dissolved. Erlenmeyer's flasks, holding about 150 c. c., will be found very convenient bottles for reducing the ferric oxide. The platinum foil should be carefully cleaned, and, if new, rubbed with fine sand to roughen it and remove the grease, &c.

When the ferric is reduced to ferrous oxide, which may be known by the absence of a blood-red color on testing a few drops of the solution removed by a glass rod to a watch-glass with ammonium sulphocyanide, empty one of the bottles into a large beaker, add 2 or 3 c. c. of concentrated sulphuric acid, and dilute to about 500 c. c.

Titrate with the standard permanganate solution in the same manner as for standardizing. The number of c. c. of potassium permanganate used, multiplied by the standard, gives the weight of metallic iron in the solution treated. From this calculate the per cent. of ferric oxide. The two titrations should not differ more than two-tenths of a c. c.

The weight of the ferric oxide thus obtained deducted from the weight of the combined oxides will give the weight of the alumina by difference.

PREPARATION OF THE STANDARD PERMANGANATE SOLUTION.

Dissolve 3.8 grams of pure crystals of potasium permanganate in 1 liter of distilled water, with constant agitation until all the crystals are dissolved, decant the perfectly clear solution, and keep in a glass-stoppered bottle.

There are several methods of standardizing the solution of permanganate. (See Sutton Volumetric Analysis, §§ 31, and 32, 33, 59.) Of these the method proposed by Marguerite, by means of iron, is to be recommended.

In determining the iron by Marguerite's method, the presence of hydrochloric acid must be avoided, especially if the solution is at all warm, since the permanganate under these circumstances will react upon the hydrochloric acid liberating chlorine, as shown in the following reaction:

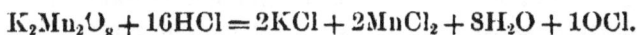

$$K_2Mn_2O_8 + 16HCl = 2KCl + 2MnCl_2 + 8H_2O + 10Cl.$$

Some of the chlorine may convert the ferrous salt present into the ferric state, but some will usually escape, and the results obtained will consequently be higher than the truth.

For this purpose introduce into a small flask, having a Kroonig valve in the stopper, 0.200 grams of cleaned piano-forte wire, containing 99.7 per cent. of iron; add 25 to 30 c. c. of dilute sulphuric acid and heat to incipient boiling. When the iron is dissolved allow the flask to cool *slowly*. When the contents of the flask are cold empty them into a large beaker, wash the flask out well, and add the washings to the main solution, dilute with distilled water to about 500 c. c.; then drop in the solution of permanganate to be standardized from a burette, having a *glass* stop-cock, with constant stirring until the color, which disappears rapidly at first and then more gradually, finally becomes permanent, and remains so for one minute. The final color should be a light pink. Note carefully the number of cubic centimeters of permanganate used, and calculate the value of 1 c. c. thus: Number of c. c. permanganate used : 1 c. c. :: weight of iron used : x; x is, therefore, equal to the value of 1 c. c. Multiply the result so obtained by 0.997 the weight of the metallic iron contained in the wire.

The standardizing of the permanganate should be repeated once or twice, and the quantity of permanganate used in the several trials should not differ by more that one-tenth of a c. c. The average may be taken as correct. (Fres., § 112, 2, *a aa.* Crookes, p. 73.)

DETERMINATION OF THE LIME.

The filtrate and washings from the hydrates are concentrated to 100 c. c., if possible, and 1. c. c. of ammonia added. If the ammonia produces a precipitate other than aluminium and ferric hydrates, which must be filtered out and added to the main precipitate of the hydrates, acidify the solution with hydrochloric acid, boil for a minute, and then make alkaline again with ammonia. This is done to introduce a sufficient amount of ammonium chloride to prevent the precipitation of the magnesium hydrate. Then add 40 c. c. of a solution of ammonium oxalate, saturated, enough to precipitate all the lime as oxalate, and convert the magnesia also into oxalate ; a very soluble compound of magnesia, and easily washed from calcium oxalate, which remains in solution. (Fres., p. 831. Exp. 92, 93.) Heat the solution to incipient boiling, and then allow it to stand undisturbed several hours. After the precipitate has settled perfectly, decant the clear fluid through a filter, wash by decantation *once* with about 25 c. c. hot water. Then dissolve the precipitate of calcium oxalate, mixed with a little magnesium oxalate, in the beaker, with as little hot dilute hydrochloric acid as possible. Should any of the precipitate have passed over on the filter, wash it back into the acid solution, dilute with about 50 c. c. of hot water, make alkaline with ammonia, and add 5 or 6 c. c. ammonium oxalate, stir, and allow the precipitate to settle. When it has settled completely, filter through the same filter into a fresh beaker, transfer the precipitate, and wash it thoroughly with hot water. The water required to transfer the precipitate to the filter will wash it sufficiently. Dry the filter and contents at a temperature not exceeding 100° C. to avoid making the filter brittle. When the precipitate is dry brush it into a clock-glass, cleaning the filter as thoroughly as possible. Burn the paper in a weighed platinum crucible until only a white ash is left. Then cool the crucible, transfer the precipitate from the glass to the crucible, add enough concentrated sulphuric acid to cover the precipitate, place the lid on the crucible, and apply a gentle heat to its edge until all the free sulphuric acid is expelled. Then ignite strongly for a few minutes, cool in a desiccator and weigh. This weight, less that of the crucible and filter ash, will give that of the calcium sulphate.

Never attempt to ignite the calcium oxalate before adding the sulphuric acid, as the ignition will convert it into calcium oxide or carbonate which effervesces violently when the acid is added, thus causing a loss of some of the substance.

It is a good plan to add to the weighed precipitate more sulphuric acid and proceed as before, until you get a constant weight. The weight of the calcium sulphate found multiplied by 0.41158 will give the weight of the calcium oxide, lime, and this divided by the weight of the soil taken for analysis and multiplied by 100 will give the percentage of lime present in the soil. (Fres., § 154, 6, *a*; and § 103, *b*, *a*.)

DETERMINATION OF THE MAGNESIA.

Make the combined filtrates from the calcium oxalate alkaline with ammonia, if not already so; add 30 c. c. of a solution of hydro-di-sodic phosphate. Agitate the contents of the beaker with a glass rod, taking care not to rub the sides, as it will cause crystals of ammonium magnesium phosphate to adhere to the sides very difficult to remove. Allow the solution to stand for twelve hours in a cool place. Filter off the clear fluid through a weighed Gooch crucible;[*] transfer the precipitate and wash with dilute ammonia, prepared by mixing one part of the strong ammonia with three parts of distilled water. In case any particles of the precipitate adhere to the sides of the beaker, rub them off as much as possible with a rubber-covered glass rod, and wash them onto the main precipitate in the crucible. Whatever cannot be so detached are moistened with a few drops of acetic acid, transferred to a smaller beaker, made alkaline with ammonia, and set aside for six hours for the precipitate to settle, and it is then added to the main precipitate.

The precipitate is washed well with the ammonical water, the crucible and contents ignited gently at first and afterwards over a blast lamp. By the action of the heat the ammonium magnesium phosphate is transformed into magnesium pyrophosphate, $Mg_2P_2O_7$. Cool the crucible and contents in a desiccator and weigh. The increase in the weight of the crucible represents the weight of the magnesium pyrophosphate; this multiplied by 0.36024 will give the weight of magnesia, MgO, present, from whence the percentage is readily calculated. (Fres., § 154, 6, *a*. § 104, 2.)

SEPARATION OF THE ALKALIES FROM THE OTHER BASES PRESENT.

In 200 c. c. of the solution B, equal to 4 grams of the air-dried soil, proceed to determine the potash and soda, in duplicate, as follows: Evaporate each 100 c. c. of the solution nearly to dryness in order to drive off as much free acid as possible; then dilute with about 75 c. c. of warm water and heat on the steam bath for half an hour. Add ammonia till the solution is nearly neutralized and then 25 c. c. of a saturated solution of barium hydrate, so that the fluid is strongly alkaline

---

[*] Proc. Amer. Acad. Arts and Sciences, 1878, p. 342.

to test paper; boil, allow the precipitate settle, decant the clear fluid on a filter, wash with 50 c. c. of hot water, by decantation, then transfer the precipitate to the filter and wash well with hot water, until all the chlorides are removed; testing a few drops of the wash-water with argentic nitrate.

Evaporate the filtrate to about 75 c. c., and add 25 c. c. of a saturated solution of ammonium carbonate, sufficient to precipitate the excess of barium present; boil, allow the precipitate to settle, decant the clear fluid, wash by decantation with 50 c. c. hot water, transfer the precipitate to a filter, and wash with hot water until all the chlorides are removed.

Evaporate this filtrate to dryness in a platinum dish, and, when dry, drive off the ammonium chloride at a low red heat. Cool, take up with water, filter, and wash well to remove the carbonaceous matter, and test the fluid with a few drops of the barium hydrate solution; if this produces a precipitate, add 10 c. c. more, or until the barium produces no further precipitate; filter off the precipitate, and repeat the treatment with ammonium carbonate.

Finally evaporate the solution to dryness in a weighed platinum dish after adding a few drops of hydrochloric acid, expel any ammonium chloride present at a low red heat, cool, and weigh the chlorides of potassium and sodium.

DETERMINATION OF THE POTASH.

Dissolve the mixed chlorides in 25 c. c. of warm water, filter, if necessary, and transfer to a small lipped porcelain dish, add 2 c. c. of dilute hydrochloric acid and 8 c. c. of platinum tetrachloride solution, prepared by dissolving one part by weight of the platinum tetrachloride in ten parts of distilled water, and evaporate to a pasty consistency on the water-bath. Then pour into the dish about 50 c. c. of 85 per cent. alcohol, without removing the dish from the bath, and heat for two or three minutes. Care must be taken that there are no ammoniacal fumes floating about in the air of the laboratory, as they would form a precipitate with the platinum tetrachloride, thus increasing and vitiating the result. Allow the precipitate to settle, and the fluid shows by its yellow color that a sufficient amount of the platinum tetrachloride has been used; decant the clear fluid through a weighed Gooch crucible, transfer the precipitate to the crucible, and wash well with 85 per cent. alcohol. (Crookes, p. 2.)

Dry the crucible and contents in the air-bath at 100° C., cool, and weigh the potassium platinum chloride. This weight multiplied by 0.30559 will give the weight of potassium chloride present. The weight of the potassium chloride is deducted from the weight of the mixed chlorides, leaving the weight of the sodium chloride present. (Fres., § 152, 1a, §§ 97, 98.)

The weight of the potassium chloride multiplied by 0.63190 will give that of the potash, K_2O, in the 2 grams of air-dried soil.

The weight of the sodium chloride present multiplied by 0.53043 will give that of the soda, Na_2O, present in 2 grams of air-dried soil.

The weights of the potassium platinum chlorides found in the duplicate analysis should agree to a tenth of a milligram.

Care must be taken that the barium hydrate used does not contain either of the two alkalies, potassium or sodium.

DETERMINATION OF THE SODA.

The soda may be determined directly, instead of by difference, as follows: Evaporate the filtrate and washings from the precipitate of potassium platinum chloride to dryness on the water-bath, and when dry burn at a low red heat. The solution contains the sodium platinum chloride, with the excess of the platinum tetrachloride used. A mixture of platinum and of sodium chloride is thus obtained; on dissolving in warm water and filtering, the sodium chloride present is washed out. The filtrate is evaporated to dryness on the water-bath in a weighed platinum dish, dried at 100° C. in an air-bath, cooled in a desiccator, and weighed. The increase in weight is due to the sodium chloride present in the 2 grams of air-dried soil. This is calculated to soda, as above.

DETERMINATION OF THE SULPHURIC ACID.

In the remaining 100 c. c. of solution B, equal to 2 grams of the air-dried soil, the sulphuric acid is determined as follows: Heat the solution to boiling, and add 10 c. c. of barium chloride, prepared by dissolving 1 part by weight of barium chloride in 10 parts of distilled water, and continue boiling three or four minutes. Allow the precipitate to settle, decant the clear fluid on a weighed Gooch crucible, pour 50 c. c. of boiling water on the precipitate, allow to settle, and decant as before. Finally transfer the precipitate of barium sulphate to the crucible, and wash well with hot water. Dry, ignite strongly, cool, and weigh. The increase in weight is due to the barium sulphate. This weight multiplied by 0.34331 will give the weight of the sulphuric acid, SO_3, present in the 2 grams of air-dried soil. (Fres., § 132, I.)

TREATMENT OF THE SOIL WITH NITRIC ACID (Sp. gr. 1.20).

Treat 10 grams of the air-dried soil, previously burnt to destroy organic matter, with 200 c. c. of nitric acid, sp. gr. 1.2, in a porcelain dish, heated to 100° C. on the steam bath for five days, and proceed in the same manner as already given in the treatment with hydrochloric acid, p. 35, to separate the insoluble residue, and then the soluble silica, taking care to wash the residue well with hot water.

Make the filtrate from the soluble silica up to 500 c. c.

DETERMINATION OF THE PHOSPHORIC ACID.

The following solutions are used in the determination of phosphoric acid; 1, ammonium molybdate; 2, acid ammonium nitrate, and 3 magnesia mixture.

PREPARATION OF THE AMMONIUM MOLYBDATE SOLUTION.

Dissolve 75 grams of ammonium molybdate in 500 c. c. of distilled water, adding the water in small quantities at a time, and filter into 500 c. c. of nitric acid of a specific gravity 1.20. One c. c. of this solution is equivalent to 0.001 grams of phosphoric acid.

PREPARATION OF THE ACID AMMONIUM NITRATE.

Add to 325 c. c. of nitric acid, specific gravity 1.2, 200 c. c. of a mixture of equal parts of ammonia, specific gravity 0.96, and water. Allow to cool, and keep in a glass-stoppered bottle.

PREPARATION OF THE MAGNESIA MIXTURE.

Dissolve 125 grams of crystallized magnesium sulphate and 125 grams of ammonium chloride in 1 liter of water; when all is dissolved add 500 c. c. of ammonia, specific gravity 0.96.

The determination of phosphoric acid is made in duplicate.

Add to 200 c. c. of the solution, equal to 4 grams of the air-dried soil, 50 c. c. of ammonium molybdate and 10 c. c. of acid ammonium nitrate, and heat on the water-bath at 80° C., with frequent stirring for four hours. Allow the precipitate to settle, and decant the clear fluid on a small filter, wash the precipitate with about 25 c. c. of acid ammonium nitrate, decant the clear fluid, and then transfer the precipitate to the filter, washing it with the same.

Set the filtrate aside for twelve hours in a warm place, after adding 10 c. c. of ammonium molybdate to insure the precipitation of all the phosphoric acid. If a precipitate should occur, which rarely happens, filter it off and add to the main precipitate.

Dissolve the precipitate through the filter, into the same beaker, with warmed dilute ammonia, 1 of ammonia to 3 of water, and as there often remains upon the filter a small quantity of iron, arising from the phosphate of that metal which the dilute nitric acid has disolved, repass the ammoniacal liquid through the filter several times. To the solution is added enough hydrochloric acid to make it decidedly acid, and then enough ammonia to render it decidedly alkaline and to redissolve the precipitate formed, and finally 5 c. c. of magnesia mixture. The latter is not to be added, however, until the solution becomes cold. After adding the magnesia mixture set the solution aside in a cool place for twelve hours, to allow the crystalline precipitate of ammonium magnesium phosphate to thoroughly settle. Filter by decantation, through a weighed Gooch crucible, wash with dilute ammonia, and proceed as in the determination of magnesia, p. 41.

The increase in weight represents that of the magnesium pyrophosphate. This weight, multiplied by 0.63976, will give the weight of the phosphoric acid in the 4 grams of air-dried soil. (Fres., § 134, I, b, β, a.)

Fresenius advises adding to the weight of the magnesium pyrophosphate 0.0018 grams, to compensate for the loss which results from the feeble solubility of the ammonium magnesium phosphate in the wash waters.

DETERMINATION OF CHLORINE.

Wash 10 grams of the air-dried soil on a filter, with boiling water, using about 500 c. c. before testing a few drops of the wash water, acidulated with nitric acid, with argentic nitrate to see that all the chlorides are washed out.

When all the chlorides are removed, concentrate the washings to 200 c. c., filter if necessary, and divide into two equal parts, determining the chlorine volumetrically by means of a standard solution of argentic nitrate, using potassium chromate as an indicator. The determination is made by first adding to the 100 c. c. of solution, equal to 5 grams of air-dried soil, 3 drops of a saturated solution of potassium chromate, and then dropping in the silver solution from a burette, and noting when the red color of silver chromate appears. The number of c. c. of silver nitrate solution used, multiplied by the value of 1 c. c., will give the amount of chlorine present in the 5 grams of soil. (Fres., § 141, b, a.)

PREPARATION OF THE ARGENTIC NITRATE SOLUTION.

Dissolve 8.5 grams of pure argentic nitrate in 1 liter of distilled water; 1 c. c. of the solution is equal to 0.001775 grams Cl. To standardize the solution, dissolve 1 gram of pure fused sodium chloride in 1 liter of distilled water, take exactly 10 c. c. of the solution and dilute to 100 c. c. with water, add 3 drops of a satuated solution of potassium chromate, and drop in from a burette the silver solution until the red color of silver chromate appears. The known quantity of chlorine in the 10 c. c. of salt solution, divided by the number of c. c. of silver solution used, will give the value of 1 c. c. of the latter.

DETERMINATION OF THE CARBONIC ACID BY ABSORPTION.

For this purpose an absorption apparatus, such as that described by Fresenius, under the head of carbonic acid, may be used. (Fres., § 139 c, § 182.)

This consists of: (1) A small tube filled with fused chloride of calcium, to absorb the atmospheric moisture; this is connected by means of rubber tubing to (2) the closed funnel, provided with a stop-cock, through which the dilute hydrochloric acid is admitted to (3) the flask, holding about 150 c. c. (4) The flask is connected by rubber tubing to an U-tube, filled with pumice stone, boiled in concentrated sulphuric acid, together with some sulphuric acid. This absorbs any moisture that may be driven off by the heat. (5) This connects again with another U-tube, $\frac{7}{8}$ of which is filled with granulated soda lime, the remaining $\frac{1}{8}$ in the upper part of the second limb contains chloride of calcium. (6) An aspirator completes the apparatus.

For the analysis, weigh the absorption tube 5, closed at both ends with small pieces of glass rods in rubber tubing, introduce about 5 to 10 grams of the air-dried soil into the decomposing flask, put the apparatus together, having previously tested it to see that it does not leak, close the stop-cock of the funnel tube, and attach the aspirator. The substance can best be weighed in a small piece of glass tubing sealed at one end, into which the soil is placed, and the weight noted; then, by shaking as much of the substance as possible into the flask, and again reweighing, the difference between this weight and the former will represent the amount of the soil taken for the determination.

When the aspirator has produced a partial vacuum, introduce the 30 c. c., about, of dilute hydrochloric acid contained in the funnel into the decomposing flask. As soon as all the acid is in close the stop-cock and apply a gentle heat until the liquid begins to boil. Then remove the heat, attach the guard tube 1, open the stop-cock, and draw air through the apparatus slowly until the liquid is cool; about 2 liters is sufficient.

Weigh the absorption tube; the difference between this weight and the first weight of the tube is equivalent to the carbonic acid contained in the quantity of soil taken.

DETERMINATION OF NITROGEN BY COMBUSTION WITH SODA LIME.

To determine the total nitrogen present in the soil proceed as follows: Select a tube of hard glass, 15 to 18 inches long, draw one end of it to a fine point, and to the other end fit tightly a cork, through which is passed a tube bent at right angles, the other end of which passes through a cork closing tightly one arm of a bulbed U-tube. Into the combustion tube first slip a loosely-fitting plug of asbestus, previously ignited, and then some 3 or 4 inches of dry soda-lime. Weigh out 1 gram of the air-dried soil, and mix it in a porcelain mortar with some finely pulverized soda-lime, and introduce the mixture into the combustion tube, forcible pressure being carefully avoided. The mixture is followed by a layer of the soda-lime, used to rinse the mortar. Enough granulated soda-lime is then added to fill the tube to 1 or 2 inches of the open end; place another plug of ignited asbestus at the end, and close with the cork carrying the tube. A free passage is formed for the evolved gases by a few gentle taps, and it is then ready to be placed in the combustion furnace, after first ascertaining that the apparatus is air-tight.

Introduce from a burette into the U-tube 10 c c. of fifth normal oxalic acid, equal to 0.028 gram of nitrogen. This is prepared by dissolving 12.6

grams of crystallized oxalic acid in 1 liter of distilled water. Add 5 c. c. of cochineal solution as an indicator. Prepared by grinding to a fine powder about 3 grams of good cochineal and macerating it with frequent shakings with 250 c. c. of a mixture of 4 volumes of distilled water and 1 volume of alcohol, 95 per cent., and filtering through Swedish paper, and keep the solution in closed bottles.

Introduce the prepared combustion tube into the furnace, letting the open end project a little so as not to burn the cork, supporting the U-tube by a clamp. The tube is then gradually heated, commencing at the fore part, nearest the cork, and progressing slowly towards the tail. Care must be taken to keep the fore part of the tube at a moderate red heat throughout the process. Avoid heating the end that is drawn to a point lest the internal pressure causes it to blow out. The combustion should be conducted so as to obtain a steady and uninterrupted flow of gas. When the tube is ignited throughout its whole length and the evolution of the gas has ceased, attach the aspirator to the other limb of the U-tube and start it slowly. Then break off the point of the combustion tube; at the same time put out the gas. Draw a slow current of air through the apparatus for a few minutes, in order to sweep all the rest of the ammonia into the acid.

Remove the combustion tube, together with the U-tube, from the furnace when the aspiration is completed, and break the combustion tube close to the cork by allowing a fine stream of cold water to fall on it. Remove the corks from the U-tube, and run in from another burette, holding fifth normal soda solution, and determine how much of the fifth normal oxalic acid solution used has been neutralized by the ammonia thus obtained from the soil. From the data thus obtained the percentage of nitrogen contained in the soil may be calculated. (Fres., § 185.)

DETERMINATION OF NITRIC ACID BY SCHLOESING'S METHOD.

The following is the method adopted at the Rothamsted Laboratory by Mr. R. Warington : *

It is very important that the sample of soil should be immediately dried when received in the laboratory, as, if this is not done, the quantity of nitric acid found may greatly exceed that existing in the original soil, as nitrification will be continually in progress whilst the soil remains damp. The temperature at which the soil is dried has a marked effect on the result. If a wet soil be dried in an air-bath at 100° C. the nitrates present will be more or less destroyed, whilst drying by mere exposure to the air is equally likely, in the case of surface soils at least, to occasion a gain in nitrates.

The following course has been adopted at Rothamsted :

PREPARATION OF THE SAMPLE.

The soil is broken up immediately it is received from the field, and spread in trays, in layers about 1 inch in thickness; the trays are then placed in a stoveroom, kept at about 55° C.; the drying is usually completed in twenty-four hours. As the temperature of the room is one at which nitrification by an organized ferment does not occur, it is probable that very little production of nitric acid takes place during the operation. After drying, stones and roots are removed, and the soil is finely powdered and placed in bottles. Soil samples thus prepared are not absolutely dry, but the small amount of water present is apparently insufficient to allow of organic change.

PREPARATION OF THE WATERY EXTRACT.

From 200 to 500 grams of the dry powdered soil are taken, according to the supposed richness of the soil in nitrates, and introduced on a large filter fitted for filtration under pressure. (Fres., § 53, a.) The filter is previously moistened, the dry soil introduced, and if the latter be of a loose texture, it is shaken firmly together, but with a clay soil this is better avoided. The flask is connected with the air-pump, the soil is kept moistened with water, and when 100 c. c. have run through, it may be concluded that all the nitrates are washed out. This operation lasts from ten to forty-five minutes, depending upon the nature of the soil.

The watery extract thus obtained is placed in a small porcelain dish, the flask washed well with water, and evaporated nearly to dryness on the water-bath. As soon as cool, it is mixed with 1 c. c. of a cold saturated solution of ferrous chloride and 1 c. c. of hydrochloric acid, both reagents having been boiled and cooled immediately before use. The mixture is then ready to be introduced into the retort.

DESCRIPTION OF THE MODIFIED SCHLOESING'S APPARATUS.

The apparatus consists of a bulb retort 1¾ inches in diameter, the tubular of which has been bent near its extremity to make a convenient juncture with the delivery tube, which dips into a trough of mercury on the left; the long supply tube attached to the receiver is of small bore, and is easily filled by a half c. c. of liquid. The short tube in the cork of the retort is also of small bore, and is connected by a piece of rubber tubing fitted with a clamp to an apparatus for the continuous production of pure carbonic acid. A long funnel tube, likewise passing through the cork of the retort, completes the apparatus.

The apparatus for the generation of the carbonic acid used is so arranged that the marble, which must previously be well boiled in water to remove as much of the oxygen present as possible, is contained in a lower reservoir into which the hydrochloric acid used, previously boiled, is introduced from a higher reservoir; thus the former is always under internal pressure, and leakage of air from without cannot occur. The hydrochloric acid, after it has been well boiled, has dissolved in it a moderate quantity of cuprous chloride, and is then introduced into the upper reservoir and covered with a layer of oil.

The presence of the cuprous chloride insures the removal of any dissolved oxygen, and gives an indication by its change of color when this condition is exceeded; as long as it remains of an olive-green tint oxygen will be absent, but should the acid become of a clear blue-green, it is no longer certainly free from oxygen, and more cuprous chloride must be added.

The reagents used must be freshly boiled and employed in as small a quantity as possible. In boiling the hydrochloric acid it is well to add a few drops of ferrous chloride, to be sure of removing any dissolved oxygen.

The mode of conducting the operation is as follows:

The apparatus previously described is fitted together, the long funnel tube attached to the bulb retort being filled with water. Connection is made with the glass stop-cock of the carbonic acid generator by means of a short, stout rubber tube, provided with a pinch-cock. The pinch-cock being opened, the stop-cock is turned till a moderate stream of bubbles rises in the mercury trough; the stop-cock is left in this position, and the admission of gas is afterwards controlled by the pinch-cock, pressure on which allows a few bubbles to pass at a time. The heated chloride of calcium bath is next raised, so that the bulb retort is almost submerged; the temperature, shown by a thermometer which forms part of the apparatus, should be 130° to 140° C. By boiling small quantities of water or hydrochloric acid in the bulb retort in a stream of carbonic acid the air present is expelled; the supply of carbonic acid must be stopped before the boiling has ceased, so as to leave little of this gas in the retort. Previous to very delicate experiments it is advisable to introduce through the funnel tube a small quantity of nitre, ferrous chloride, and hydrochloric acid, rinsing the tube with the latter reagent; any trace of oxygen remaining in the apparatus is then consumed by the nitric oxide formed, and after boiling to dryness, and driving out the nitric oxide with carbonic acid, the apparatus is in a perfect condition for a quantitative experiment.

The mixture of the extract with ferrous chloride and hydrochloric acid is introduced through the funnel tube, and rinsed in with three or four successive half cubic centimeters of hydrochloric acid. The contents of the retort is then boiled to dryness, a little carbonic acid being from time to time admitted, and a more considerable quantity used at the end to expel any remaining nitric oxide.

The gas is collected in a small jar over mercury. The gas analysis is of a simple character; the gas is measured after absorption of the carbonic acid by potash, and again after absorption of the nitric oxide, the difference giving the amount of this gas. For the absorption of nitric oxide, a saturated solution of ferrous chloride was for some time employed. This method is not, however, perfectly satisfactory when the highest accuracy is required, the nitric oxide being generally rather underestimated, except the process of absorption is repeated with a fresh portion of ferrous chloride. The error is greater in proportion to the quantity of unabsorbed gas present. The use of ferrous chloride as an absorbent for nitric oxide has now been given up, and the oxygen method substituted. All the measurements of the gas are now made without shifting the laboratory vessel; the conditions are thus favorable to extreme accuracy.

The chief source of error attending the oxygen process lies in the small quantity of carbonic acid produced during the absorption with pyrogallol; this error becomes negligible if the oxygen is only used in small excess. The difficulty of using the oxygen in nicely regulated quantity may be removed by the use of Prof. G. Bischof's recently invented "gas-delivery tube." This may be made of a test tube, having a small perforation half an inch from the mouth. The tube is partly filled with oxygen over mercury, and its mouth is then closed by a finely perforated stopper, made from a piece of wide tube, and fitted tightly into the test tube by means of a covering of rubber. When this tube is inclined, the side perforation being downward, the oxygen is discharged in small bubbles from the perforated stopper, while mercury enters through the side opening. Using this tube, the supply of oxygen is perfectly under control, and can be stopped as soon as a fresh bubble ceases to produce a red tinge in the laboratory vessel. The trials made with this apparatus have been very satisfactory.

REMARKS.

Where such a complete analysis of a soil is not required, as that for which the directions are given in the preceding pages, the estimation of potash, soda, phosphoric acid, nitrogen, and lime will give valuable information for judging of its fertility.

The following qualitative tests may be applied in case only a very preliminary examination is required.

Test the slightly moistened soil with litmus paper; if this should show an acid reaction, the presence of an excess of humic acids, or small quantities of sulphate of iron, may be suspected. All good and fertile soils have generally no effect on litmus paper, or show only a slight alkaline reaction.

Make a water solution of the soil, and test the solution for lime with ammonium oxalate; for sulphuric acid with barium chloride, after acidulating with hydrochloric acid; for iron, with ammonia; and for chlorides with silver nitrate. An excess of any of the three latter would indicate that the soil contains injurious quantities of them.

Boil some of the soil with nitric acid, and after filtering off the insoluble residue, test the solution with ammonium molybdate for the presence of phosphoric acid.

In a hydrochloric acid solution of the soil the different bases may be tested for as in the quantitative analysis. If an effervescence is produced on adding the acid to the soil, the presence of carbonate of lime is indicated, but should none occur, but analysis show that lime is present, it is probably in the state of sulphate or gypsum.

ON THE GEOLOGICAL CHARACTER AND DISTRIBUTION OF THE SOILS IN THE UNITED STATES.

While there is a vast variety of detail in the character of the soils of this country in regard to both their physical properties and chemical composition, still they may be classified under the two heads of soils of transport and soils of disintegration, geologically speaking.

Soils of transport include, as has been previously stated, all drift and alluvial materials which have been worn from other rocks by atmospheric agencies and transported to their existing positions by ancient glacial action, by winds, and by waters. These embrace the majority of all soils occurring in the United States.

Drift soils.—These occupy the principal portion of the States lying north of the Ohio and east of the Missouri Rivers. According to Professor Dana, they occur " over all New England and Long Island, New York, New Jersey, and part of Pennsylvania, and the States west, to the western limits of Iowa and Minnesota. Beyond the meridian of 98° W., in the United States, they are not known. They have their southern limit near the parallel of 39°, in Southern Pennsylvania, Ohio, Indiana, Illinois, and Iowa, whilst their northern is undetermined. South of the Ohio River they are hardly traceable." *

Without going into the details of the theory of ancient glacial action, which has given rise to a large amount of study and an extensive literature, the term *drift*, as it is commonly employed in geology, includes the sands, gravels, clays of various composition and texture, and bowlders, more or less water-worn, all mingled in various proportions and of various degrees of fineness, which have been transported from places in higher latitudes by glacial action and deposited on the country rock in varying thickness.

The soils of this drift are usually gravelly, often stony, of variable fertility, from the noted fertile lands of Ohio and Western New York to the barren portions of New England. As a whole, these soils grow finer as they go further southward and westward from New England and Western New York. When overcropped and worn out, as often happens, they recover when allowed to rest fallow several years by the decomposition of the mingled materials of which they are composed.

Alluvial soils.—These are formed from the deposits of the fine earthy materials, sediment, silt, or detritus, by running streams and rivers, of which we have such a notable example at the Mississippi's Delta.

The amount of transportation going on over a continent is beyond calculation; streams are everywhere at work; rivers, with their large tributaries and their thousand little ones, spreading among all the hills and to the summits of every mountain. And thus the whole surface of a continent is on the move toward the oceans. The word *detritus* means worn out, and is well applied to river depositions. The *amount*

* Dana's Geology, p. 528.

of silt carried to the Mexican Gulf by the Mississippi, according to the Delta Survey under Humphreys and Abbot, is about $\frac{1}{1500}$ of the weight of the water, or $\frac{1}{2900}$ of its bulk; equivalent for an average year to 812,500,000,000,000 pounds, or to a mass 1 square mile in area and 241 feet deep.[*]

These constitute the "bottom lands," as they are called in the West. The Red River region, which has become famous as a wheat producing country, lying partly in Minnesota and partly in Dakota, occupies the bed of an ancient lake, known to geologists as Lake Agassiz, and is composed of a black sedimentary soil, exceedingly fine in texture, and very fertile and deep. This tract extends southward to Lake Traverse, on the Red River, widening as it proceeds northward and extending on both sides of the river 50 or 60 miles wide where its bed leaves this country, and expanding to much greater width in Manitoba.

The further westward soils of this class are found the less the amount of organic matter they contain, although the soils are not necessarily less fertile, until in some places in the valleys of California are found soils of great fertility which contain an exceedingly small amount. Of course such soils, as those of California just mentioned, are deficient in the faculty of storing up water for future use, and, however rich they may be in mineral constituents, yet in a dry region or one subject to periodical droughts, irrigation would have to be resorted to in order to get large yields of crops.

Soils of disintegration.—These occupy the undulating parts of this country lying south of the drift, possessing every variety of character, both in regard to their chemical composition and physical properties, as their mode of formation indicates, arising from the disintegration of the subjacent rocks by atmospheric agencies.

Where the underlying rock has been an impure limestone, containing much insoluble matter, the carbonate of lime has been slowly dissolved out by the action of the carbonic acid contained in the rain, leaving the insoluble matter behind. Such soils as that of the "blue grass" regions of Kentucky are so formed, and are often of extreme fertility. (See the "Kentucky Geological Reports" for further details about this region, including the chemical analyses of its soils.)

Professor Whitney states that some of the prairie soils of Iowa, particularly those where the soil is almost of impalpable fineness, have been produced by the slow action of atmospheric agencies on beds of limestones which formerly occupied their places. In the course of time the soluble carbonate of lime was gradually dissolved out and carried away by the rivers and streams to the ocean, and a small amount of insoluble residue was left, forming the thick prairie soil of the region, which has since become blackened by the decay of subsequent abundant vegetation on it.[†]

In the table-lands of Oregon and Washington the underlying rock is volcanic, and the soil arising from its disintegration is very fine in texture,

---

[*] Dana's Geology, p. 648.
[†] Iowa Geological Survey, Vol. I, 1858.

dark in color, of great fertility, and, judging from the soils of similar origin found in the Rhine region and the Mediterraneans in Europe, which have supported vineyards for many years, will probably prove very enduring and produce a great variety of crops.

These two classes of soils run into each other by insensible gradations.

The term " prairie soils " is most indefinite, as commonly used, including soils of various origin. The prairie region of the West occupies a vast extent of country, extending over the eastern part of Ohio, Indiana, the southern portions of Michigan and Wisconsin, nearly the whole of Illinois and Iowa, and the northern portion of Missouri, and gradually passing, in Kansas and Nebraska, into the *plains*, or the arid and desert region which lies at the base of the Rocky Mountains. West of the parallel of 97° and 100° the country becomes too barren to be inhabited and worthless for cultivation.

The region of the greatest cereal production of this country includes the most noted of the prairie soils, and is nominally in the drift region of geologists. Light clays and heavy loams are the best for wheat, though very heavy clays often produce good crops, both as to yield and quality; the lighter soils may yield a good quality, but deficient in quantity; moderately stiff soils produce generally the best crops.

HISTORY OF THE SOILS ANALYZED BY THIS DIVISION.

During the past year over thirty-six soils were analyzed by this division, thirty of which were done completely and the results obtained will be found in Table IV. The remainder were only partially analyzed, and are not tabulated.

This table is presented in the following pages, and the history of each soil, as far as known, is appended, to be found under its respective serial number.

TABLE IV.—*Analyses of air-dried soils, by Edgar Richards.*

	Prairie soils from Dakota.			Hon. Joseph Jorgensen, United States land office, Walla Walla, Wash.							N. E. Smith.
Soil received from											
Soil marked.	A. No. 6.	B. No. 16.	C. No. 7.	Sandy soil from 5 miles northwest of Umatilla, Oreg.	Surface in Grant's Ranch, S. 24, T. 11, R. 24.	Two feet of surface in Grant's Ranch, S. 24, T. 11, R. 25.	T. 8, R. 26.	S. 26, T. 7, R. 26.	Between Yakima and Columbia Rivers, middle of T. 8 N., R. 27.	S. 12, T. 8, R. 28.	Union Pier, Berrien County, Michigan.
Serial number	1611.	1612.	1913.	1056.	1657.	1658.	1659.	1660.	1661.	1662.	2550.
Percentage of—											
Moisture	6.275	7.800	7.700	.525	1.300	1.950	1.600	.675	1.325	1.125	8.300
Insoluble silica	69.335	53.415	39.555	78.602	62.165	62.640	63.640	71.585	67.575	64.860	38.935
Hydrated silica		13.620	21.215	5.983	17.600	16.485	16.105	11.460	13.925	16.185	7.575
Soluble silica	.490	.460	.485	.275	.260	.275	.470	.370	.575	.385	.135
Sesquioxide of iron, Fe₂O₃	4.096	4.008	3.204	3.920	4.800	5.248	5.056	4.256	4.736	4.768	.960
Alumina, Al₂O₃	7.952	9.930	7.382	4.008	6.738	6.818	5.740	3.828	5.510	6.238	3.775
Phosphoric acid, P₂O₅	.112	.112	.224	.192	.192	.224	.224	.216	.224	.224	.715
Lime, CaO	.848	.852	3.898	1.338	1.433	1.329	2.099	1.418	1.428	1.449	.907
Magnesia, MgO	.868	1.535	2.007	.703	.650	.465	1.411	.973	.047	.991	.207
Potash, K₂O	.720	.725	.745	.440	.495	.475	.545	.535	.940	.700	.180
Soda, Na₂O	.945	.640	1.550	1.690	1.560	1.070	.830	.945	1.255	.700	.344
Sulphuric acid, SO₃	.120	.077	.103	.043	.052	.080	.069	.052	.085	.129	.275
Chlorine, Cl	.027	.053	.078	.020	.015	.020	.030	.020	.007	.014	.249
Carbonic acid, CO₂	.220	.104	2.530	.005	.002	.116	.110	.000	.116	.000	1.380
Volatile and organic matter	8.905	6.171	10.175	2.045	3.573	3.584	2.040	1.885	1.559	2.600	36.186
Total	100.913	100.102	100.971	100.479	100.844	100.759	99.969	99.828	100.157	100.368	100.272
Nitrogen, N	.324	.179	.414	.041	.089	.075	.069	.140	.224	.067	1.148
Air-dried soil contains—											
Coarse gravel											34.92
Fine material											65.08

TABLE IV.—*Analyses of air-dried soils, by Edgar Richards*—Continued.

Soil received from	Jesse H. Blair.		William Cartwright, Oswego, N. Y.									
Soil marked.	Lebanon, Boone County, Indiana.		South field.				North field.			Hart's field.	
	Soil.	Subsoil.	Southeast corner, A. 1.	Southwest corner, A. 2.	Northeast corner, A. 3.	Northwest corner, A. 4.	Southeast corner, B. 5.	Northeast corner, B. 6.	Northwest corner, B. 7.	East end, C. 8.	West end, C. 9.
Serial number	2851.	2852.	2853.	2554.	2555.	2556.	2557.	2558.	2559.	2560.	2561.
Percentage of—											
Moisture	5.300	3.975	3.950	3.025	5.475	1.825	5.595	5.890	1.440	2.885	1.850
Insoluble silica	58.175	62.050	78.195	80.965	78.560	82.250	60.135	61.575	74.160	70.035	69.615
Hydrated silica	10.970	12.965	3.095	4.490	3.210	4.720	5.975	9.025	8.165	8.505	9.520
Soluble silica	.290	.215	.190	.205	.495	.155	.335	.177	.170	.120	.177
Sesquioxide of iron, Fe_2O_3	.272	2.720	2.432	2.308	1.952	2.592	2.076	3.360	3.200	3.206	3.360
Alumina, Al_2O_3	6.397	7.583	4.615	3.740	2.885	3.368	5.234	5.872	4.922	5.468	5.544
Phosphoric acid, P_2O_5	.041	.127	.023	.052	.023	.050	.010	.048	.038	.176	.176
Lime, CaO	1.387	1.280	.350	.440	.350	.564	.753	1.873	.535	.683	.634
Magnesia, MgO	.771	.872	.389	.501	.274	.641	.555	.668	.642	.793	.746
Potash, K_2O	.510	.575	.320	.425	.305	.595	.400	.475	.480	.530	.460
Soda, Na_2O	.725	.280	.165	.215	.220	.240	.990	1.010	1.370	.830	.880
Sulphuric acid, SO_3	.300	.223	.172	.138	.180	.206	.318	.103	.069	.086	.086
Chlorine, Cl	.014	.011	.011	.011	.014	.007	.014	.011	.014	.018	.025
Carbonic acid, CO_2	1.260	1.104	.653	.180	.166	.102	.302	1.283	.605	1.239	.699
Volatile and organic matter	12.165	5.221	5.622	3.370	6.134	2.023	7.794	8.617	4.645	5.511	6.551
Total	100.577	100.101	100.482	100.125	100.252	100.238	100.320	100.187	100.435	100.175	100.323
Nitrogen, N	.574	.252	.156	.101	.162	.109	.221	.313	.153	.204	.218
Air-dried soil contains—											
Coarse gravel	9.64	23.53	32.04	31.86	19.40	30.00	37.73	39.00	48.10	36.00	46.72
Fine material	90.36	76.47	67.96	68.14	80.60	70.00	62.27	61.00	51.90	64.00	53.28

TABLE IV.—Analyses of air-dried soils, by Edgar Richards—Continued.

Soil received from	F. Soip, Alexandria, Rapides Parish, Louisiana.			Mrs. William Waters, Alexandria, La.			William Harris, Alexandria, La.	
Soil marked.	Alluvial new land, one year cleared, No. 1.	Alluvial medium or "chocolate," twenty years in cultivation, No. 2.	Alluvial red clay, thirty years in culti No. 3.	Alluvial "bottom land," front, and sandy, fifty years in cultivation, No. 4.	"Creek bottom, low land, sixteen years in cultivation, gr. label No. 1.	"Pinchill land," gr. label No. 2.	Alluvial "bottom land," front and sandy, wh. label A.	Alluvial "bottom land," front, and sandy, wh. label B.
Serial number	2574.	2575.	2576.	2577.	2579.	2580.	2581.	2582.
Percentage of—								
Moisture	4.235	2.000	2.650	.675	1.625	.375	.700	.900
Insoluble silica	47.351	62.968	46.679	77.920	75.647	91.157	83.854	81.590
Hydrated silica	21.231	15.817	22.391	9.060	9.498	3.045	6.313	7.002
Soluble silica	.155	.101	.065	.035	.030	.02)	.030	.040
Sesquioxide of iron, Fe_2O_3	4.384	3.200	4.544	2.240	1.440	.800	1.760	1.728
Alumina, Al_2O_3	10.090	7.156	10.213	4.127	5.333	1.840	3.030	3.246
Phosphoric acid, P_2O_5	.160	.144	.193	.113	.097	.000	.080	.006
Lime, CaO	1.165	2.000	.836	.414	.185	.111	.371	.026
Magnesia, MgO	2.169	1.066	2.547	1.131	.346	.090	.839	1.934
Potash, K_2O	1.470	.930	1.940	.805	.430	.165	.745	.805
Soda, Na_2O	.780	.700	.915	.730	.745	.460	.650	.730
Sulphuric acid, SO_3	.080	.049	.052	.052	.052	.027	.054	.080
Chlorine, Cl	.014	.011	.014	.018	.032	.032	.014	.025
Carbonic acid, CO_2	.540	1.330	1.713	.940	.763	.308	.275	.905
Volatile and organic matter	6.451	2.639	3.462	1.735	4.187	2.142	1.650	.520
Total	100.296	100.260	100.214	100.015	100.410	100.572	100.365	100.533
Nitrogen, N	.209	.120	.137	.078	.140	.059	.073	.073
Air-dried soil contains—								
Coarse gravel	.72	.00	.00	.00	.00	5.00	.00	.50
Fine material	99.28	100.00	100.00	100.00	100.00	95.00	100.00	99.50

LABORATORY, UNITED STATES DEPARTMENT OF AGRICULTURE.

PRAIRIE SOILS FROM DAKOTA.

1611–1613. These soils were forwarded to the Department in July, 1882, unaccompanied by any letter or other means of indentification from the person who sent them; their analysis was begun in expectation that some information concerning them would come to hand before they were finished, but all attempts to find out the sender have, so far, proved unavailing.

SOILS FROM THE UNITED STATES LAND OFFICE, WALLA WALLA, WASH.

The seven samples of soil were sent by Hon. Joseph Jorgenson, United States land office, Walla Walla, January 5, 1884.

They were taken from various points of a section of unsettled country, lying between the Yakima and Columbia Rivers, and west of Wallula, on the Northern Pacific Railroad, comprising about 1,300 square miles of gently rolling plateau—from 500 to 1,000 feet above sea level—the only drawback being a lack of running streams of water on any part of it, and but few natural springs. Water is reached at varying depths, from 14 to 80 feet. It is covered, however, with a fine bunch grass, which is accepted here as indubitable proof that the smaller grains will grow to maturity and perfection. This year (1885) there are some fine crops of wheat on it.

The samples were taken from "1 to 5 feet" in depth, the soil being a "decomposed basalt from 3 to 100 feet deep," and the subsoil is "basaltic rock." No timber is found on it, the prevailing growth being "bunch grass and sage bush."

1656. Sandy soil from 5 miles northwest of Umatilla, Oreg.

1657. Surface soil in Grant's Ranch, Sec. 24, T. 11, R. 24.

1658. Two feet of surface soil in Grant's Ranch, Sec. 24, T. 11. R. 25.

1659. Soil from T. 8, R. 26.

1660. Soil from Sec. 26, T. 7, R. 26.

1661. Soil from middle of T. 8 N., R. 27, between the Yakima and Columbia Rivers.

1662. Soil from Sec. 12, T. 8, R. 28.

These are samples of virgin soils, and contain a large amount of the most important soil constituents, as phosphoric acid, lime, potash, &c., and should produce abundant crops under favorable climatic conditions. In their contents of nitrogen, however, they are, with the exception of Nos. 1660 and 1661, somewhat deficient; and this would indicate that ammoniacal manures would have to be applied in the future, if, by excessive cropping, the soil should become unproductive.

SOIL FROM N. E. SMITH, UNION PIER, BERRIEN COUNTY, MICHIGAN.

2550. The sample of soil was sent by Mr. Smith December 10, 1883. The sample was taken to a depth of "10 inches from a portion of the inverted furrow." The field is "flat" and the depth of the soil "like sample is from 8 to 30 inches." The subsoil, "to a depth of 2 feet, is sand filled with the infiltration of the surface; this sand in places has many small flat stones resembling pieces of broken oyster shells in shape but flinty in character." The timber was "yellow pine and larch, filled with a dense growth of alders, tag and black, and blueberries; the surface was covered with moss 2 feet deep." The following crops have been raised:

Oats, good straw, light grain. Buckwheat, 25 to 30 bushels to the acre. Corn, not a success. Potatoes, one hundred and one in a hill, but none larger than a walnut. Cabbages, radishes, melons, squashes, and beans have succeeded. My largest experience is with onions from the seed; the first year, after getting 2 inches high, many turned yellow on top and finally died; second year they were better, and third year good.

In regard to manure used:

In plats as follows: First year, ashes and lime, fresh slaked; ashes 200 bushels to the acre, lime 2 tons to the acre; crop failed. Same plat, second year: Hen droppings at the rate of 10 cubic yards per acre, composited with plaster, and just previous to application mixed with twice their bulk of white-ash ashes. Yield, 300 to 400 bushels per acre. Third year: "Garden City phosphate," 1,000 pounds per acre. Yield improved. This year (1885) applied nitrate of soda, 150 pounds, "Garden City phosphate," 800 pounds per acre; the crop is of fair promise in the main, but there are spots where a good stand has disappeared; in these barren spots there will be found small patches of fine onions marking the spot of a fire. The original plat, treated this year as above, now (July) promises a fine crop. This year I have taken in new ground with the above stated result.

The sample was dried to make it more secure when sent through the mail.

This sample, as the most casual inspection of the analysis will show, contains an enormous amount of organic matter, and to this may be attributed the poor success met with in raising crops; as nothing is more injurious than the action of the organic acids, arising from the decay of the organic matter in the soil, on vegetation when they are present in excess. For, however fertile the soil may be in other respects, until this excess of humic acids is neutralized or otherwise got rid of, the prospect of raising remunerative crops is very slight. The remedy for such a state of affairs is a heavy dressing of lime, from 2 to 5 tons of quicklime per acre, depending on the quantity of organic matter present, that is, from 0.05 to 0.5 per cent., by weight, of the cultivated soil; the lime or marl used has the power of neutralizing the humic acids. Burning might also be resorted to, but the use of lime will probably, in such cases, prove more beneficial. The lime should be used as a top dressing, as it has a strong tendency to sink into the subsoil, and so it should not be plowed in, but kept as near the surface as possible. The ground should be plowed first, then the lime spread and simply harrowed in. This dose of lime must not be repeated yearly, but at intervals of six or eight years 1 to 2 tons of lime made into a compost may be used. It is best applied in the early winter, so that the lime may work into the surface before the spring growth commences.

The amount of nitrogen and of phosphoric acid is very large, and that of lime, of potash, and soda is abundant for the raising of any crop when the excess of organic acids has been destroyed. With the exception noted the analysis shows this soil to be a very fertile one, containing an abundant supply of all the necessary plant constituents.

SOIL AND SUBSOIL FROM JESSE H. BLAIR, LEBANON, BOONE COUNTY, INDIANA.

2551, 2552. The samples were sent January 5, 1884, having been taken on September 12, 1883, from "what is popularly called a prairie region, but what is thought to have once been a lake, in the northern part of Hendricks County; it was dry and very difficult to get a good sample." The sample of soil was taken "by digging a hole an inch square, then shaving a slice downward, about 6 inches deep." The sample of subsoil was taken from the "next 6 inches below the surface sample." The soil is "rich, solid, and about 18 inches deep, and in a meadow of timothy grass." The subsoil is "tough clay, about 3 feet deep, then sand or gravel." "No timber, a swamp or wet prairie, and lately redeemed." No manure has been used. The following crops were raised: "Corn, 75 bushels per acre; large yield of broom corn, then a large yield of hay." "It produces a heavy crop of grass; wheat does fair; the corn is not as good as clay lands yield."

The analyses show that an abundant supply of the necessary plant constituents are present, and that the soil should be very fertile. The amount of nitrogen in the soil is very large.

SOILS FROM WILLIAM CARTWRIGHT, OSWEGO, N. Y.

2553–2561. Samples taken from three distinct fields on which an acre of sugar-beet was grown in 1883, and were sent December 24, 1883. Samples Nos. 2553–2556, marked "A 1, 2, 3, and 4," were taken from "a square two-thirds acre plot at different points, SE., SW., NE., NW. of the field." Samples Nos. 2557–2559, marked "B. 5, 6, and 7," were from "a triangular one-third acre plot," taken at the different angles. The two remaining samples, Nos. 2560 and 2561, marked "C. 8, and 9," were from "a field of sugar-beet a mile distant" from the other two fields, "cultivated by another party, on a rectangular plot of half an acre; the samples being taken at the ends, E. and W., of the rectangle."

The general character of all the fields was a gentle slope, enough to turn water readily. The samples were cut out with a spade, a couple of weeks after the crop was gathered, each about 6 inches wide and deep; the soil of field A was 8 to 10 inches deep; that of field B probably 1 foot; field C was rather stony, soil 8 to 12 inches deep The subsoil of all the fields was hard-pan, with large stones and bowlders imbedded. A subsoil plow was used in preparing fields A and B. No timber was grown on the fields; the woods adjacent, I believe, were maple. The land had been under cultivation for years. Fields A and B had been heavily manured in the spring of 1882 with barn-yard manure, and an excellent crop of corn and beans gathered that year. A succession of rotating crops had been taken previously from these two fields, but I have not the statistics concerning them. No manure was directly applied previous to beet planting on A and B, but I was informed that on field C barn-yard manure was strewn midway between the beet rows, which were 30 inches apart. In fields A and B, after harrowing and rolling, the seed, sugar-beet seed was sown, part by hand and part with a wheelbarrow drill, in rows 18 and 20 inches apart, on the 4th and 9th of May, 1883. All work after hoeing, thinning, and weeding was entirely by hand. The crop weighed nearly 18 tons.

The analysis of the beets grown on these different fields is as follows:*

Analyses of beets from William Cartwright, Oswego, N. Y.

Variety.	Number of analyses.	Number of beets taken.	Total weight.	Weight without neck.	Per cent. sucrose.	Per cent. glucose.	Per cent. ash.	Coefficient of purity.
			*Kilos.**	*Kilos.**				
Improved, south field, north end...	1	5	2.838	Not taken.	12.12	0.29	1.022	74.6
Improved, south field, south end...	4	5	2.457	2.238	15.34	0.17	0.775	83.0
Improved, north field, north end...	5	5	2.776	2.610	15.32	0.16	0.862	85.0
Improved, north field, south end ..	6	5	2.795	2.540	15.20	0.12	1.061	82.0
From Hart's field.................	11	5	2.915	2.810	12.74	0.40	0.897	79.0

*A kilogram is equal to 2.2 pounds.

The analyses of these soils show the great difficulty of obtaining a sample of soil from a field which shall represent its average quality, unless the greatest care is taken.

In regard to the analysis, No. 2553–2556, taken from the south field at different corners of the plot, the three samples, A 1, 2, and 4, contain practically the same amount of coarse sand and gravel, whilst A 3 has about 10 per cent. less. All four samples show that the soil is deficient in phosphoric acid and lime, and probably would be much benefited by the use of a lime phosphate or similar fertilizer; its contents of other soil constituents are ample for fertility.

The samples, No. 2557–2559, taken from the north field, show that this soil is likewise deficient in phosphoric acid, but is richer in its contents of lime and nitrogen and in other constituents similar to that of the south field. The content of gravel also varies in the different samples.

The two samples, No. 2560 and 2561, taken from Hart's field, differ in their contents of coarse gravel, but contain an abundance of phosphoric acid and other soil constituents.

For the purpose of comparing soils on which such sugar-producing plants as sorghum and sugar beet have been grown, the following analyses, made by Mr. Clifford Richardson, first assistant chemist of this Department, in 1882–'83, are given:†

ANALYSES OF SOILS.

The character and composition of the soils best adapted to the cultivation of sorghum for sugar production, as, also, the proper method of fertilization necessary for the best results, are obviously matters of fundamental importance.

At present our knowledge is very limited, and the number of carefully ascertained facts so small as hardly to warrant more than conjecture.

In many respects the habits of the sorghums and their demands upon climate and soil are almost identical with those of the several varieties of maize, and yet there appear to be in certain respects marked differences. It is known that when fairly

* Chemical Division, Bulletin No. 3, 1884, p. 26.
† Investigation of Sorghum as a Sugar-producing Plant; season of 1882. Special Report, pp. 58–64.

established the sorghums as a class are capable of sustaining a period of drought which would prove fatal to maize, and not only this, but that such drought and the accompanying high temperature results in the development of an unusual amount of sugar in the plant. (See Annual Report of Department of Agriculture, 1881-'82, p. 456.)

It will be seen by consulting the results of our experiments as to the effect of fertilizers upon the sugar content and ash in the juices of the several sorghums (see Annual Report, 1880, pp. 118, 125) that, although a very large number of determinations were made, the average result of all was such as to leave the matter wholly unsettled.

To those who may desire to aid in these and similar investigations, a careful study of these results above referred to may be helpful as showing the extreme danger of hasty generalizations; for any half dozen of the analytical results, selected at random and considered alone, would, in most cases, warrant a conclusion, more or less decided, which the increase of testimony renders less and less probable.

The results of the past year at Rio Grande, N. J. (where they produced 320,000 pounds of sugar, and where, upon fields identical in character, there was great variation in the amount of crop produced), were such as to awaken great interest in the questions of soils and fertilization. Besides, the juices of the sorghums there grown proved to be remarkably pure, comparing well even with the best sugar-cane juice. Therefore, average specimens of the soils from the several fields were obtained, and a record of the yield of crop and the fertilizers applied to each was also secured from the president of the Sorghum Sugar Company, George C. Potts, Esq., of Philadelphia, Pa.

Rio Grande is a small hamlet some 6 miles north of Cape May, N. J., in latitude 39 degrees north and longitude nearly 2 degrees east from Washington. It is situated upon a sandy peninsula, about 5 miles in breadth, with the Atlantic upon the east and separated from the mainland by the Delaware Bay, at this point about 2 miles wide. Average samples of soil from six fields were selected for analysis, viz:

A. *Harne farm.*—This field received an application of 300 pounds of Peruvian guano per acre. The average yield of stalks was $3\frac{1}{4}$ tons per acre.

B. *Richwine farm.*—This farm also had 300 pounds Peruvian guano per acre. The average yield was $5\frac{1}{4}$ tons of stalks per acre.

C. *Hand farm.*—This field received an application of 300 pounds of Peruvian guano and 30 bushels of lime per acre. The average yield was $7\frac{1}{4}$ tons of stalks per acre.

D. *Neafie farm.*—This field received same amount of guano and lime as C. Average yield per acre, 8 tons stalks.

E. *Uriah Creese farm.*—Same amount of guano and lime as C and D. Average yield per acre, 15 tons stalks.

F. *Bennett farm.*—Same amount of guano and lime as C, D, and E. Average yield per acre, 17 tons stalks.

From the above results it will be seen that the application of the expensive fertilizer Peruvian guano was without any apparent benefit, while the application of lime seems to have been beneficial, although it is to be regretted that we have not the data for comparing the yield of these fields with and without the application of fertilizers.

With the exception only that the amount of pebbles of an appreciable size, one-twentieth to one-quarter inch in diameter, was more in some of the samples than in others, there was to the eye no noticeable difference in the character of the six.

The samples were passed through sieves of 20, 30, 40, 50, 60, 70, 80, 90 meshes to the inch, and the following results obtained: The column marked residue consisted of pebbles which would not pass through a sieve of twenty meshes to the inch, or rather of one-twentieth inch diameter. The column marked 20 was that portion which, passing meshes of one-twentieth inch, would not pass those of one-thirtieth, &c.

Besides these six samples of soil from Rio Grande, N. J., analyses have been made of several other soils upon which sorghum was grown the past year, as follows:

G. *Grounds of the Department of Agriculture.*—The recent treatment of this plot is given in the annual reports of the past three years. The sample for analysis was taken November, 1882.

H. *Soil No. 1—Great Bend, Kans.*—This soil has been cultivated for six years. The yield was 10½ tons stalks per acre. No fertilizer used.

I. *Soil No. 2—Great Bend, Kans.*—This soil was plowed for the first time. The yield per acre was 8 tons of stalks. No fertilizers were used.

J. *Soils from Rising City, Nebr.*, upon which 18 tons per acre of sugar-beets were grown, which gave, on analysis, an average of 12·27 per cent. of sugar in the juice.

K. *Soil from Hutchinson, Kans.*—Yield of sorghum, 6 tons stalks per acre.

L. *Soil from Sterling, Kans.*—Under cultivation for three years in cereal crops. A black, sandy loam. Average yield per acre, 7 tons stalks.

M. *Soil from Sterling, Kans.*—A black, sandy loam. Under cultivation for seven years with crops of cereals. Crop very promising, but destroyed by hail.

N. *Soil from Sterling, Kans.*—Black, sandy loam. Under cultivation for five years in cereal crops. Average yield per acre, 12 tons of stalks.

O. *Soil from Sterling, Kans.*—A strictly sandy soil; in cereals for five years. Average yield per acre, 10 tons of stalks.

Per cent. of soils passed through sieves.

	Residue.	20.	30.	40.	50.	60.	70.	80.	90.	Total.
A	27.8	4.2	5.9	3.6	2.5	3.0	4.3	3.9	44.8	100
B	22.7	5.1	8.9	7.8	5.0	6.0	11.2	6.4	26.9	100
C	3.0	6.7	17.6	16.5	8.8	8.8	11.4	10.6	16.6	100
D	5.7	7.2	16.7	13.6	9.6	8.2	9.8	12.0	17.2	100
E	8.2	6.2	12.2	8.7	7.5	6.9	9.8	8.3	32.3	100
F	5.5	6.9	18.6	12.5	8.7	9.4	9.7	8.0	20.7	100
G	5.5	1.6	6.6	7.2	3.9	3.6	5.1	7.9	58.6	100
H	2.2	1.1	2.9	3.3	3.7	2.6	7.3	6.8	70.1	100
I	0.3	0.3	0.9	1.0	1.2	0.7	2.2	4.5	88.9	100
J	0.2	0.5	0.1	0.1	0.1	0.6	1.1	1.5	95.8	100
K	0.7	0.3	0.8	1.0	0.7	0.9	2.1	2.0	91.6	100
L	1.8	3.0	7.4	5.6	5.1	3.1	4.6	4.6	64.8	100
M	4.9	1.4	3.3	3.2	2.1	2.7	4.2	3.3	74.9	100
N	1.0	0.9	3.3	7.5	6.5	4.2	9.6	12.9	54.1	100
O	0.9	1.6	9.2	8.6	11.3	9.3	14.6	14.7	29.8	100

So far as the partial mechanical analysis goes it quite fails to throw any light upon the cause of the very wide difference in the crops grown upon the Rio Grande soils.

For example, the soils C, D, F are very much alike, and yet their respective yields per acre in tons of stalks were 7½, 8, and 17. It is obvious that much of this might have been due to difference in cultivation, but it does not appear that there was practically any difference in this respect.

CHEMICAL COMPOSITION OF THE SOILS.

The following table shows the results of the chemical analysis of the several soils. The absence of other than mere traces of chlorine in the Rio Grande soils is remarkable, in view of the fact that these fields were lying within a few hundred yards of the ocean. It is possible that the heavy fall rains had leached such compounds below the surface, from which alone the samples were taken for analysis. It is intended to make still further examination of the subsoils of these several fields, for it may be that sorghum, being through its root system a deep feeder, will account for good crops of cane upon land which failed to grow good crops of other kinds:

Percentage of—	A.	B.	C.	D.	E.	F.	G.
Water	.830	.680	.190	.350	.430	.180	1.140
Organic matter	4.730	3.500	1.290	2.180	2.420	1.780	4.690
Insoluble matter	87.008	92.243	96.910	93.897	93.167	95.297	84.670
Fe_2O_3	2.555	1.775	.940	1.110	1.500	1.445	3.440
Al_2O_3	4.110	1.610	.550	1.765	1.805	1.060	4.360
CaO	.315	.305	.225	.375	.460	.505	.860
MgO	.390	.290	.147	.185	.180	.190	.367
K_2O	.238	.124	.061	.085	.122	.074	.394
Na_2O	Trace.	.023	Trace.	....	.012	Trace.	.023
P_2O_5	.088	.047	.024	.034	.043	.026	.265
SO_3	Trace.	Trace.	Trace.	Trace.	Trace.	Trace.	Trace.
Cl	.044	.009	.003	.004	.005	.003	.009
CO_2	....	....	....	.130	....	Trace.	Trace.
	100.308	100.636	100.340	100.055	100.144	100.560	100.228
Nitrogen	.128	.067	.045	.045	.078	.062	.146

Percentage of—	H.	I.	J.	K.	L.	M.	N.	O.
Water	1.000	.300	1.140	.330	.400	.470	.300	.360
Organic matter	4.320	5.820	7.320	4.830	4.310	5.150	2.520	1.330
Insoluble matter	85.250	84.625	78.162	86.282	87.792	81.832	91.544	94.231
Fe_2O_3	3.005	3.330	4.550	3.270	2.775	3.270	2.330	1.775
Al_2O_3	3.575	3.890	5.805	3.385	3.005	3.665	1.835	1.465
CaO	.710	.760	.715	.565	.660	2.685	.450	.505
MgO	.325	.450	.820	.595	.380	.690	.390	.230
K_2O	.524	.538	.086	.437	.482	.397	.301	.257
Na_2O	....	....	Trace.	.059	Trace.	.042	.050	Trace.
P_2O_5	.047	.046	.042	.026	.010	.017	.024	.017
SO_3	Trace.	.115	Trace.	.050	Trace.	.044	.036	.041
Cl	.004	.019	.017	.007	.027	.019	.019	.017
CO_2	....	....	....	....	....	1.695	....	....
	99.360	99.893	99.257	99.836	99.941	99.976	99.790	100.226
Nitrogen	.151	.190	.230	.162	.140	.146	.034	.050

For purpose of comparison, analyses are given of two sugar-cane soils from a pamphlet on the agricultural chemistry of the sugar cane by Dr. T. L. Phipson.

A is soil from Jamaica, under cane for the first time.

B is soil from Demerara which has been steadily under cane for 15 years.

	A.	B.
	Per cent.	Per cent.
Moisture	12.25	18.72
Organic matter and combined water	15.36	6.03
Silica and insoluble	48.45	68.89
Alumina	13.80	2.50
Oxide of iron	6.72	2.60
Lime	.99	.08
Magnesia	.29	.25
Potash	.11	.10
Soda	.70	.09
Phosphoric acid	.10	.03
Sulphuric acid	.30	.03
Chlorine	*.51	Trace.
Oxide manganese, carbonic acid, and loss in analysis	.42	.68
	100.00	100.00
Nitrogen in organic matter	.31	.05

* This quantity of chlorine is unusually high, and is accounted for by the proximity of a salt spring.

Dr. Phipson calls attention to the greater amount of organic matter, nitrogen, lime, and phosphoric acid in A, and to the important fact that the quantity of lime, .08, in B is far below that of the magnesia, .25. This he regards as a very bad sign in cane soil. He deduces from the results of a numerous series of analyses made by him that the degree of exhaustion which a cane soil has suffered may be determined by comparing the relative amounts of lime and magnesia present in them.

In support of this view, he gives analyses of four samples of cane soils from Guiana,

A and B having been cultivated from ten to fifteen years and C and D having been cultivated over sixty years:

	A.	B.	C.	D.
Lime .... per cent..	.44	.64	.11	.40
Magnesia ....do....	.32	.50	.36	.51

In view of the above facts, it is not improbable that a similar explanation will suffice for the remarkable results obtained at Rio Grande, N. J.

In the following table the crop of stalks produced, with the per cents of lime and magnesia in the several soils, is given for purpose of comparison with ratio of lime to magnesia:

	Tons stalks.	Per cent. lime.	Per cent. magnesia.	Ratio lime to magnesia.	
A	3¼	.315	.300	100	124
B	5½	.305	.290	100	95
C	7	.225	.147	100	65
D	8	.375	.185	100	49
E	15	.460	.180	100	39
F	17	.505	.190	100	38

It will be remembered that while each of the above soils had received an application of 300 pounds of Peruvian guano per acre, the soils C, D, E, and F had, in addition, received 30 bushels of lime per acre. It is also very interesting to observe that as the relative amount of magnesia compared with lime in the above soils fell off the crop of cane increased.

For purposes of comparison, the tons of stalks produced per acre, with the per cents of the lime and magnesia, and their ratio, is given for the other soils analyzed:

	Tons stalks.	Lime.	Magnesia.	Ratio lime to magnesia.	
G	15	.860	.367	100	43
H	10½	.710	.325	100	46
O	10	.505	.230	100	46
L	7	.660	.380	100	58
I	8	.760	.450	100	59
N	12	.450	.390	100	87
K	6	.505	.595	100	105

In the above list the order of arrangement is according to the ratio of lime to magnesia, and it will be seen that the crop from soil N is the only one which is fairly exceptional to the conclusions laid down by Dr. Phipson in his examinations of sugar-cane soils. The ratios of L and I are almost identical, and there is but a ton difference in the yield per acre; also the actual amount of lime present in I is greater than that in L.

The results at Rio Grande, N. J., in the use of lime show the importance of determining the question as to what fertilizers are best suited for sorghum, not in increasing the crop, but in improving the quality of the juice as to content of sugar and coefficient of purity.

Especially are experiments desirable in the application of the various lime fertilizers, as superphosphates, sulphate of lime, quick lime, and powdered limestone.

SOILS FROM RAPIDES PARISH, LOUISIANA.

2574–2577. Soils from the cotton plantation of F. Seip, situated on Bayou Rapides, near Alexandria, Rapides Parish, Louisiana:

All these four samples were taken from the same plantation, and their differences simply arise from the greater or less distance from the water course in which the plantation lies; near the stream the soil is lighter or sandier; as it recedes it becomes heavier, until finally the red clay soil is reached. The land is a part of what is known as the "bottom" or alluvial lands of the Red River Valley. These lands are level, having but a slight elevation above tide water, and in their native state covered by a growth of heavy timber or forest. They lie near the Red River, and are drained by smaller streams or bayous running into the Red. The principal timber growth is sweet gum, various kinds of oak, ash, hackberry, sycamore, elm, mulberry, pecan, cotton-wood, &c. Trees are often from 3 to 6 feet in diameter, and a height of 75 feet is not uncommon. Some of the land has been recently cleared, whilst other parts have been for many, seventy or more, years in cultivation.

The samples were, in every instance, taken to a depth of 6 inches and 6 inches square, or as near that as practicable The character of the soil for some 10 feet or more is principally a red clay, with an occasional mixture of clay and sand. The surface for a few inches is a black mold, arising from the decay of vegetable matter, the leaves of the forest, &c. Beneath the red clay is generally found a blue or grayish clay.

The crops grown consist of corn and cotton, the latter principally. The yield would average in the past five years 250 pounds of lint cotton per acre; under favorable conditions of weather and good culture, 500 pounds and over were obtained. Corn would average about 25 to 30 bushels per acre. No manure was used.

2574. This sample was taken from a "field of some 8 or 10 acres but one year cleared, the remainder, 300 acres in extent, being heavily timbered, but of a similar formation."

2575. "This soil has been twenty years in cultivation and proved very fertile, and is a sample of medium or 'chocolate' land."

2576. This soil has been longer in cultivation than either of the two preceding, viz, "thirty years," and is a specimen of the fertile "red clay."

2577. This is a sample of the "front and sandy alluvial lands, and has been fifty years in cultivation, producing a somewhat smaller crop" than No. 2575.

2579-2580. Soils from Mrs. William Waters, samples collected by Mr. H. R. Cummings, Alexandria, La.:

2579. This is a sample of what is known as "creek bottom land", having been taken from "Flaggan Creek," near Alexandria, La.

The term is applied to the narrow belts of land bordering on each side of the small creeks in the Pine Hills. In this particular locality the formation extends on both sides of the creek over a thousand acres. Owing to its slight elevation the land is subject to overflow; the ground is slightly undulating, and situated within a few hundred yards of the creek, into which it easily drains. The soil is generally thin, not more than 12 inches deep. The subsoil is stiffer and soon becomes a thick blnish clay, intermingled with sand and gravel. The principal forest growths are white oak, hickory, beech, ash, and magnolia.

This sample was taken from a "field of 20 acres, which has been sixteen years in cultivation, in corn, cotton, and oats. Yield from 30 to 40 bushels of corn, and from 200 to 300 pounds of lint cotton. No manure has been used except by planting peas in the corn."

2580. This was taken from a "field in the Pine Hills, back of the 'creek lands,' and is a fair specimen of these lands, which embrace three-fourths of the area of this parish. The lands are high, rolling, and heavily timbered with pines, *Pinus palustris*, and are not much valued for cultivation. The lands being hilly are easily and naturally drained into the creeks. The field from whence the sample was taken has been cultivated in corn, cotton, and oats, with light yields. In good seasons not more than 10 to 15 bushels of corn and 100 to 125 pounds of lint cotton per acre have been produced. The soil is only a few inches deep, and the subsoil consists of sand, gravel, and clay."

2581 and 2582. Soils from the plantation of William Harris, on Bayou Robert, near Alexandria, La. :

These soils are of the same formation as those taken from Mr. Seip's plantation, and possess similar characteristics, being "alluvial bottom lands" of the Red River Valley.

In regard to the analyses, No. 2574 and 2576, the samples agree very closely in their contents of the more important soil constituents, viz, phosphoric acid, potash, lime, &c., though the amount of nitrogen in the former is nearly double that in the latter, which might be expected from a virgin soil.

No. 2575 and 2577 show a less amount of potash, phosphoric acid, and nitrogen than No. 2574, owing to their having been under cultivation for a longer period, and no attempt having been made to keep up the supply by the use of manures. As far as chemical analysis is concerned, all these soils are rich enough in all the necessary soil constituents for the continued raising of abundant crops, though the continued cropping, year after year, without the use of manure is not to be recommended if an abundant yield is to be maintained. A moderate application of farm-yard manure, or the ashes of the cotton plant and seeds mixed with lime, would certainly result in an increased yield.

The sample of "creek bottom land" No. 2579, is deficient in its contents of lime, and the application of this fertilizer would undoubtedly increase the productiveness of the land. In other respects it is sufficiently rich.

The analysis of the sample of "Pine Hill land," No. 2580, shows the complete absence of phosphoric acid, and a great deficiency of lime; in fact, it is nearly all pure quartz sand. It would seem to be a hopeless task to bring such soils to any degree of profitable fertility, as there is such a general deficiency of the most important plant constituents. The continued application of such fertilizers as South Carolina phosphates, containing both lime and phosphoric acid, and farm-yard and cotton-seed manures, with the admixture of some of the "red clay" soils, would in course of time greatly improve such lands; and as they cover nearly three fourths of the area of this parish, some such course as above indicated will have to be adopted. The mere application of lime in liberal quantities would have a beneficial effect.

The application of lime to the soils, Nos. 2581 and 2582, from Mr. William Harris, would increase their fertility, as they are somewhat deficient in their contents of lime.

C

www.ingramcontent.com/pod-product-compliance
Lightning Source LLC
Chambersburg PA
CBHW022021080426
42733CB00007B/673